THROUGH THE *electronic* LOOKING GLASS

3-D Images from a Scanning Electron Microscope

Dee Breger

CYGNUS GRAPHIC
Phoenix, Arizona, U.S.A.

Through the Electronic Looking Glass

First English/German Edition, June 1995

German translation by:
Dr. Siegmar Wittig
Wittig Books
Hückelhoven, Germany

Published by:
CYGNUS GRAPHIC
P.O. Box 32461
Phoenix, Arizona 85064
U.S.A.

Process photography:
Techniprint Co., Phoenix, Arizona, U.S.A.

Printed by:
Reliable Reproductions, Tempe, Arizona, U.S.A.

Streifzug durch den Mikrokosmos

Erste englische/deutsche Auflage, Juni 1995

Übersetzung ins Deutsche:
Dr. Siegmar Wittig
Wittig Fachbuchverlag
Hückelhoven, BRD

Verlag:
CYGNUS GRAPHIC
P.O. Box 32461
Phoenix, Arizona 85064
U.S.A.

Fototechnische Arbeiten:
Techniprint Co., Phoenix, Arizona, U.S.A.

Druck:
Reliable Reproductions, Tempe, Arizona, U.S.A.

ISBN 0-942927-90-7

Introduction

At one time or another, nearly all of us have looked through a microscope at some minute object and been amazed at what we've seen. Just as the astronomical telescope allows us to look into the universe across distances we can't even begin to imagine, the microscope lets us peer into a miniature world that is both far removed and at the same time intimately connected to the relatively enormous world we know.

The magnifiers and microscopes in use today are, of course, much more complex, vastly superior, and infintely more powerful than the instruments of earlier times; some of the newest instruments can now provide such high magnifications that individual atoms can be seen.

We're not going quite that far in this book, though. Rather, we're presenting a collection of 3-D scanning electron microscope images created by Dee Breger of Columbia University. Some of the images are of things you may not have ever seen or even heard of, but others are of things that you're very familiar with and perhaps use every day.

There's a greater difference between two-dimensional and three-dimensional images than you might realize. To illustrate this, we've included both 2-D and 3-D versions of each image. We invite you to compare what you see in each of them; by doing so, we believe you'll understand more clearly why 3-D images can be so valuable in many scientific, technological, and educational applications.

Creating 3-D imagery is a lot of extra work for the microscopist, so this type of SEM imagery has seldom been seen. Those of us without access to these research instruments are fortunate that Dee is not only an expert microscopist, but also an enthusiastic fan of stereoscopic imagery; because of this, she creates many of her images as stereo pairs instead of single 2-D images. It's this combination of expertise and interest that brings you the in-depth views of the microworld in the following pages as she sees them through her "electronic looking glass".

Duncan Woods
Cygnus Graphic

Author's Note

The micrographs in this book were originally produced on Polaroid Type 55 film with a Leica/Cambridge 250 Mark 2 scanning electron microscope at the Lamont-Doherty Earth Observatory of Columbia University, in Palisades, New York. Lamont-Doherty is a world-class earth science institute, where intense studies of our planet are conducted in the fields of continental and marine geology, paleontology, physical and chemical oceanography, biology, seismology, magnetic and gravity fields, and global climate. Scientists from Columbia and other universities and museums have used Lamont's scanning electron microscope in their research, and many of the micrographs in this book were taken from their samples. I'm greatly indebted to the researchers who donated these samples and provided the scientific information about them.

Of all the scientists who have used Lamont-Doherty's SEM, those who took advantage of its 3-D capabilities were amply rewarded; they learned more about the complicated interrelationships of their samples' ultramicroscopic features than they ever could from "flat" 2-D pictures. I expect that as older analog SEMs (such as the one that produced these stereo pairs) are replaced in the transition to contemporary digital instruments, this will encourage more use of 3-D imagery to reveal truths still held secret in the microscopic world. A few former secrets are revealed in this collection of 3-D anaglyphs.

I wish to express my deepest thanks to Duncan Woods, of Cygnus Graphic, who has made the publication of these 3-D images into a reality, and to Dr. Siegmar Wittig, of Wittig Books in Germany, for his translation of the English text into German, which will allow many more people to read and learn about the subjects of these micrographs. I hope you enjoy looking into the microworld from this unique perspective as much as I have.

Dee Breger
Lamont-Doherty Earth Observatory
June, 1995

Microscopes: optical and electronic "Looking Glasses"

If you form a small loop in one end of a blade of grass or a piece of fine wire and then dip the loop in water, you will have made the simplest and most elementary, but nevertheless quite usable, magnifier. The surface tension of the water droplet will cause it to cling to the loop in a double-convex shape, creating a tiny lens with low magnifying power.

Simple magnifying glasses, limited to about 10 times magnification, were known by the ancient Greeks and eventually came into fairly common use by the 14th century. By the 1670s, Antonie van Leeuwenhoek of Holland had developed the art of grinding and polishing tiny simple lenses of glass, clear quartz crystals, and even diamonds to such perfection that many experts believe his lenses could magnify images up to 300 times, based upon the the very accurate and detailed drawings he made of tiny microorganisms. It was not until sometime between 1590 and 1610 that the compound microscope (which has two or more lenses) came into being. Although numerous individuals had experimented with multi-lens devices, historians believe that the first practical compound microscope was invented by the Dutch lens-makers Hans and Zacharias Janssen. Even though we now look back upon these early magnifiers and microscopes as very primitive, their invention and development was still revolutionary because they permitted scientists to see, for the first time in history, the minuscule world surrounding us that is invisible to the unaided eye.

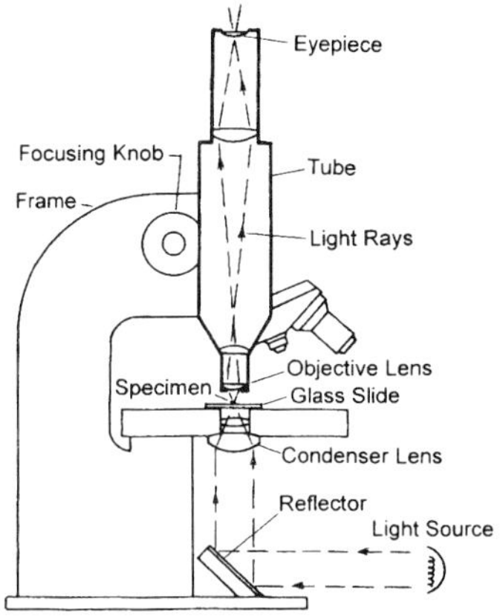

Over the next three centuries, the compound optical microscope (also called a light microscope) evolved into the instrument we know today. As shown in this diagram, the modern microscope consists basically of a number of lenses housed within a tubular structure. Light is either transmitted through or reflected from a specimen on the microscope stage up through the lens system to the eyepiece, where you can look directly at, or photograph, a greatly magnified image of the specimen. A major advantage of the optical microscope is that you can see specimens in their full, natural color, and you can examine practically anything that will fit on the stage: insects, crystals, larger bacteria, a strand of your own hair, a feather, and even the tiny living organisms in a drop of pond water.

The optical microscope has some disadvantages, though. Particles of light, or *photons*, are quite large in comparison with other types of subatomic particles, and this limits the *resolving power*, and thus the *effective magnification*, of the light microscope. Even the very best optical microscopes with lenses of the highest quality cannot effectively magnify beyond about 1,400 times without the addition of highly specialized and expensive equipment. Another disadvantage of this type of microscope is that its *depth of field* (its range of clear focus) is also very limited; if you place a specimen of some thickness on the stage, you will be able to focus clearly on only one level at a time. For example, if you focus your microscope to see the upper part of the specimen clearly, you won't be able to see a sharp image of the lower part, but if you focus the instrument to see a clear image of the lower part, then the upper part will become blurred and indistinct.

Because of the shortcomings of light microscopes, scientists began to experiment in the early 1930s with using *electrons*, rather than photons, to create images. Electrons, the negatively-charged subatomic particles responsible for electricity, are far smaller than photons, and so permit much higher resolving power and effective magnification. They can also be easily directed and brought to a focus; in this case, electromagnetic lenses are used rather than the glass lenses found in optical microscopes.

The first electron microscope was the *transmission electron microscope* (TEM) that created magnified images by passing an electron beam through a very thin specimen; by the late 1930s, a number of research laboratories were using TEMs, which could magnify specimens up to about 500,000 times. Despite the tremendous advances in microscopy brought about by the TEM, depth of field was still a problem, preparation of samples was time-consuming and difficult, and many specimens were destroyed in the process. Scientists then began to experiment with the creation of a different type of electron microscope.

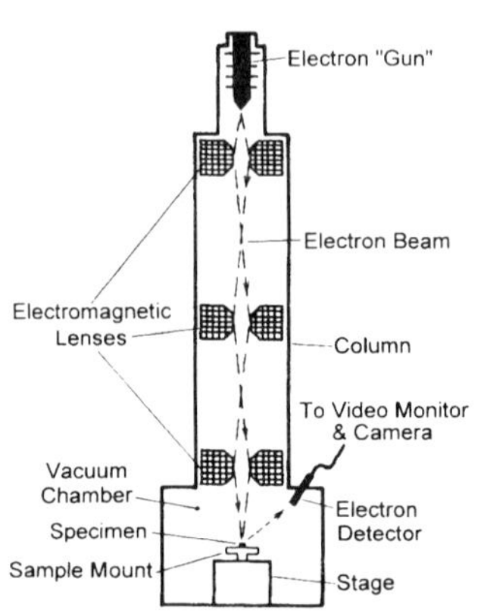

These experiments resulted in the invention of the *scanning electron microscope*, or SEM, often called simply the "s-e-m" or "sem". Development of this microscope began in 1937, but progress was very slow and the first practical commercial SEMs didn't appear until the 1960s. Despite a strong resemblance to TEMs, SEMs are quite different. This diagram shows how SEMs operate: the electron "gun" at the top of the instrument is aimed through a stack of electromagnetic lenses at the specimen below and emits a powerful beam of electrons, whose impact on the specimen causes it to emit *secondary electrons* from its own surface. These secondary electrons are then collected by a detector that converts them into a visual image on a monitor, providing an amazingly detailed portrait of the specimen.

Contemporary SEMs are capable of magnifying images 700,000 times life size (although most micrographs are taken well below this figure) and – unlike light microscopes and TEMs – they have enough depth of field to easily show many levels of even large specimens in sharp focus. By using the various controls and adjustments on the SEM's electronic console, the microscopist can scan across the entire surface of the specimen, move, rotate, and tilt it in any direction, zoom in for a close-up look at a particular detail and then zoom out again for a broader overview, or hover in one place for a longer look at an especially interesting detail or to create a micrograph (an SEM photograph) of it.

Just what does magnification in these ranges mean? Try to imagine this: if your thumbnail were enlarged 300,000 times, it would be about as large as the southern half of Manhattan Island in New York! Conversely, if this same part of Manhattan Island were reduced 300,000 times to the size of your thumbnail, you could place it in a scanning electron microscope and examine the buildings, streets, cars, and people. A magnification of 300,000 times is considered "low" by today's standards: exotic and unbelievably powerful new microscopes, including the scanning-tunneling microscope, the atomic force microscope, the acoustic microscope, the scanning laser acoustic microscope, and the transmission positron microscope, have been developed that can magnify images up to *500 million times* -- for the first time ever, we can now actually produce images of individual atoms themselves!

Electron microscopes have their own disadvantages, though. Unlike the photons of visible light, electrons cannot convey color information, so all images are produced in black and white; color, if desired, must be added by artificial means, either manually, photographically, or electronically. Additionally, it's practically impossible to view living biological specimens with conventional electron microscopes because samples must be dry and electrically conductive to withstand bombardment by the strong electron beam in the vacuum chamber. By the time they reach the microscope, most specimens are no longer living and their "natural states" are often somewhat altered. New SEMs that can view non-conductive specimens in an atmosphere of gas or liquid have been developed, but they're very expensive and still quite rare. Even with these disadvantages, though, all types of electron microscopes show us astoundingly small objects much as we would see them if it were possible to do so with the unaided eye, and they have opened a window with a view into a submicroscopic world that was unimaginable only a generation ago.

SEM images are often described as "three-dimensional" because of their solid appearance and large depth of focus. True depth, however, can be created only through stereoscopy. To make stereoscopic images, the microscopist follows the same principles and procedures used by stereo photographers, with only a few minor technical differences. For stereoscopic SEM micrographs, the microscopist produces two images of the same specimen from slightly different angles, or viewpoints. This is, in fact, how you see stereoscopically day in and day out: each of your eyes sees a slightly different view, and these two views are merged together in the brain to create the perception of depth. It is indeed fortunate that SEM technology is capable of stereo imaging, since this is the only way objects in the microworld can be seen as they really exist.

Using the Scanning Electron Microscope

Using a scanning electron microscope requires much more time, effort, training, and experience than a light microscope, particularly in the creation of 3-D images. The following brief descriptions of the various steps in the operation of an SEM will give you an idea of the complexity involved in using these sophisticated scientific instruments.

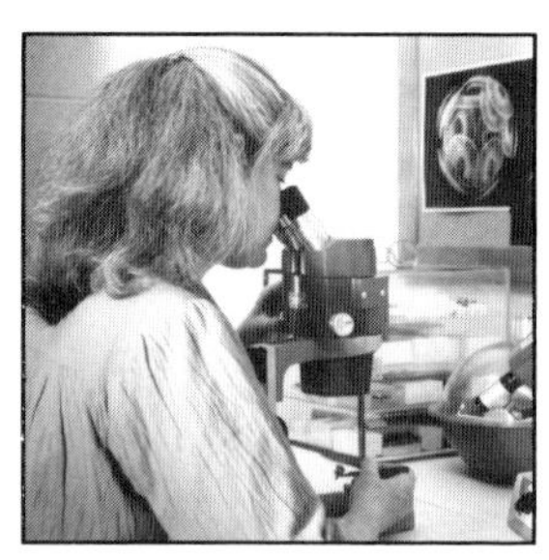

Sample preparation: Samples are first carefully affixed onto small metal specimen mounts in a variety of ways that depend on their size and shape. Some samples are large enough to be easily recognized without any magnification, while others are so small that thousands of them can fit onto the surface of a 12mm diameter mount and still not be visible until they are viewed through either a light microscope or electron microscope. Here, Dee is using a binocular light microscope to insure that a small South Pacific Ocean barnacle is securely glued to the top surface of the specimen mount.

Metal coating: Most SEMs require electrically conductive specimens. The mounted non-conductive barnacle is being placed into the vacuum chamber of a coater, where it will be bathed in metallic gas created by heating metal (in this case a gold-palladium alloy) in a very low pressure atmosphere of argon. In this process, which is called sublimation, the metal changes directly from a solid to a gas and is deposited as a very thin film onto the surface of the specimen.

Loading the SEM: The door of the sample chamber has been opened so that several mounted samples can be placed inside, after which it is closed and a vacuum is created in the chamber. The inner stage can be controlled to move, tilt, and rotate the samples as their images are displayed on the SEM's viewing monitors at any magnification from about eight times up to about 300,000 times life-size. The Polaroid camera that records the images is attached to its own screen at the upper right side of the control console.

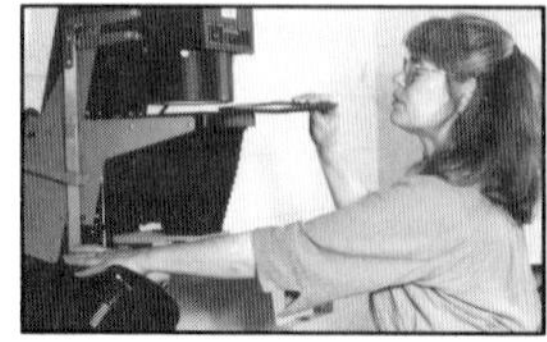

Printing the micrographs: The SEM's camera produces both positive prints and negatives. Here, a negative is being placed into a photographic enlarger that will project the image onto a sheet of photographic paper to create a positive print. The image on the print is developed with chemical baths, either manually in trays or by using automatic developing machine.

Checking the prints: After the developing chemicals are washed off, the prints are dried; several of them are shown here coming out of the automatic dryer. They're now checked again closely for tonal ranges and overall quality. Stereo pairs are more difficult to print than single micrographs because the contrast and brightness of the two images must match one another as closely as possible.

Viewing: This is perhaps the most enjoyable step in making stereo pairs. In this picture, colleague Ted Koczinsky is shown viewing a new stereo pair with a mirror stereoscope that allows the two separate 2-D prints to be seen as one 3-D image. The stereo pairs in this book were produced in the scanning electron microscope by the process described above and then photographically converted to stereo anaglyphs for this book. Turn to Plate 1 on the next page and you'll see what Ted is viewing here.

Photos: (facing page) Nancy Howard Gray; (this page) Duncan Woods.

About the 3-D Images

The stereo anaglyph has been chosen as the most suitable 3-D technique for this work. The principle is quite simple: a stereo pair (two separate images which differ slightly from one another because they've been created from different viewpoints) is first converted into lithographic halftones. These halftones are then printed in superimposition in two mutually exclusive colors, either red and blue or red and green. The printing colors are mixed to match the filters of the viewing glasses as exactly as possible. When you look at an anaglyph image through the viewer, each eye is limited to seeing only its own appropriate view, and the illusion of depth is created just as it is in your normal stereoscopic vision.

We offer the following tips for the easiest and most enjoyable viewing of the 3-D images:

Good lighting is essential to viewing anaglyphs, since the filters of the viewer absorb a lot of the light reflected from the pictures and prevent it from reaching your eyes. Viewing the images under a strong, bright light will help you see the 3-D depth much more clearly and also more quickly.

Hold the viewer with the red lens in front of your left eye. If you reverse the viewer and hold it with the blue lens before your left eye, you'll see a "pseudoscopic" image in which everything is reversed – the foreground will appear to be in the background and vice versa – and which will most often appear very illogical to the mind.

It's important that you take your time and allow each image to develop fully for the best three-dimensional effect; this is especially true if you have little or no experience in viewing 3-D imagery. View each of the images straight on from your normal reading distance for at least 30 seconds; you'll find that the longer you view each one, the more apparent and clear the depth and entire image will become.

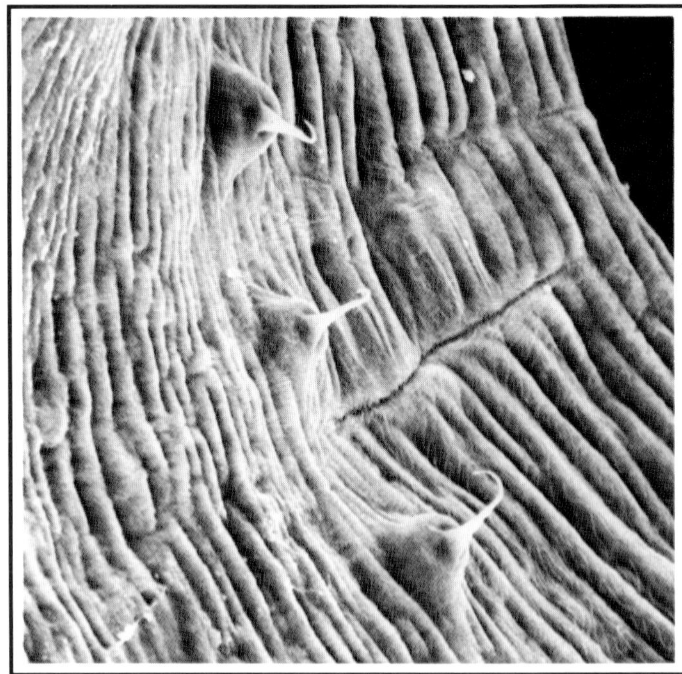

2-D: *500X*↗; 3-D: *1000X*→

Plate 1: Barnacle

This micrograph shows a surface detail on the outer shell of a juvenile shallow-water barnacle from the South Pacific Ocean. The little mounds support hollow whiskers that extend through the shell from the animal inside; the whiskers contain sensitive tissue that may serve to detect water currents or perhaps warn the barnacle of a predator climbing its shell. The many lines extending across the shell are growth ridges that generally correspond to tidal conditions in the barnacle's particular habitat.

Bild 1: Rankenfüßer

Diese Aufnahme zeigt ein Detail der äußeren Schale eines jungen, im Flachwasser des südpazifischen Ozeans lebenden Rankenfüßers. Die kleinen Erhebungen geben den kurzen, hohlen Härchen Halt, die aus dem Inneren des Tieres durch die Schale hindurchwachsen. Diese Härchen enthalten ein empfindliches Gewebe, das möglicherweise Wasserströmungen spürt oder das Tier vor Freßfeinden warnt, die die Schale berühren. Die vielen feinen Furchen in der Schale entstehen während des Wachstums und hängen von den speziellen Verhältnissen im Lebensraum des jeweiligen Tieres ab.

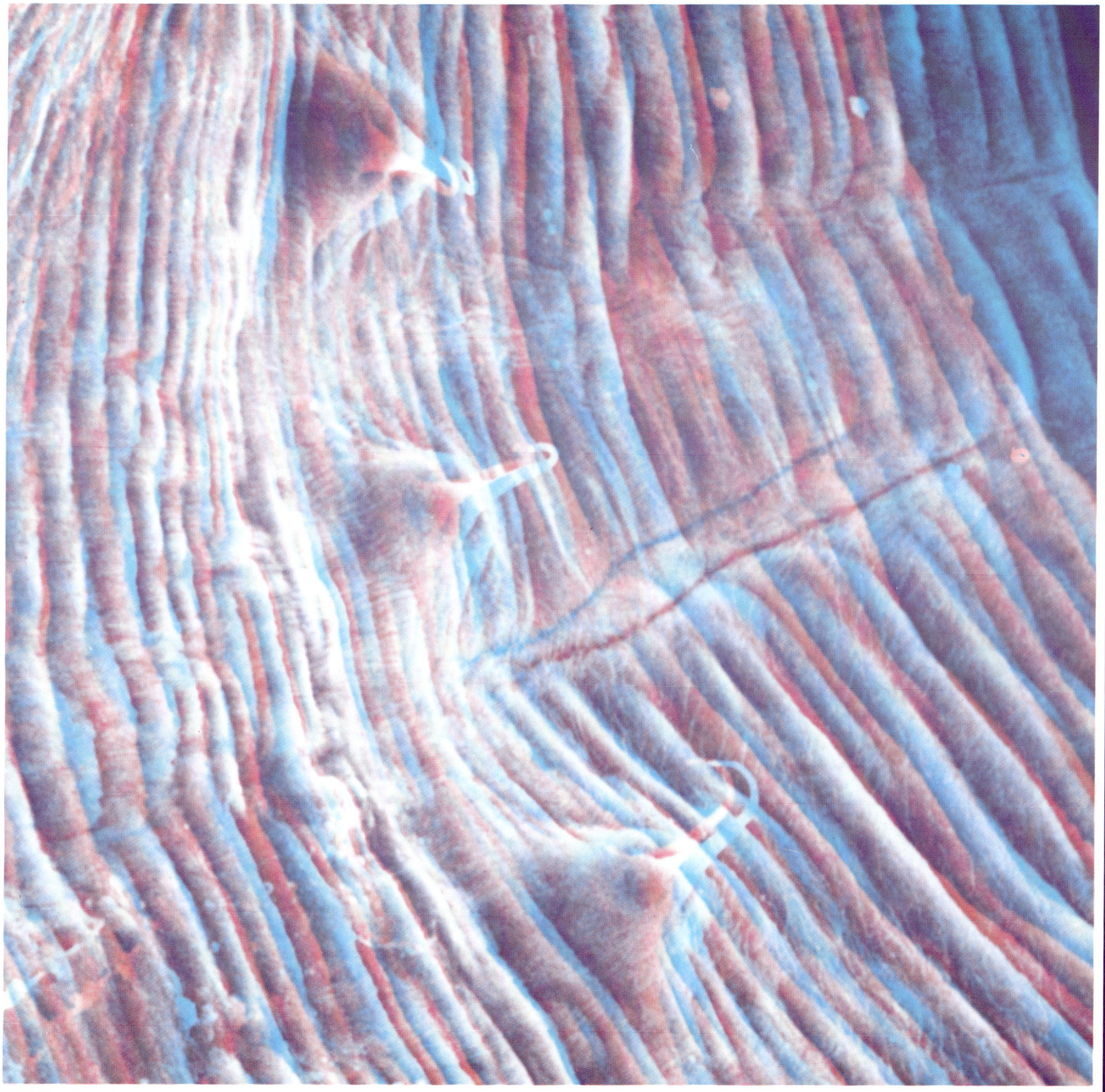

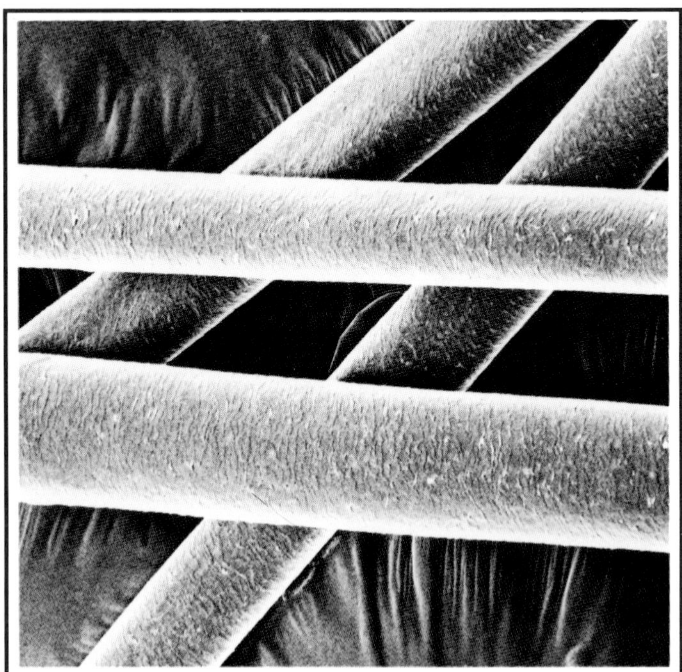

2-D: *150X*↗; 3-D: *300X*→

Plate 2: Hair

Hair is made of keratin, the same versatile material of which fingernails, reptile scales, and insect bodies are made. A strand of hair grows only within the sack-like follicle at its base, where cell division occurs. As it grows out of the follicle, densely compacted dead protein cells form an outer layer of thin overlapping scales that surrounds a core of fibers; many of these scales can be clearly seen in this micrograph of four strands of human hair. Examination of hair is often useful in diagnosing disorders of the endocrine system.

Bild 2: Haare

Haare bestehen aus Keratin (Hornstoff), dem gleichen, vielseitigen Materiel, aus dem auch Fingernägel, Insektenkörper und die Schuppen der Reptilienhaut aufgebaut sind. Ein Haar wächst nur im säckchenförmigen Haarbalg an seiner Wurzel, wo Zellteilung auftreten kann. Während es aus dem Haarbalg herauswächst, bilden tote Eiweißzellen eine dichte, äußere Schicht von dünnen, sich überlappenden Schuppen, die die inneren Fasern ummanteln. Viele dieser Schuppen sind auf dem Bild, das vier Haare zeigt, deutlich zu sehen. Eine solche Untersuchung der Haare ist oft nützlich bei der Diagnose von Funktionsstörungen innerer Drüsen.

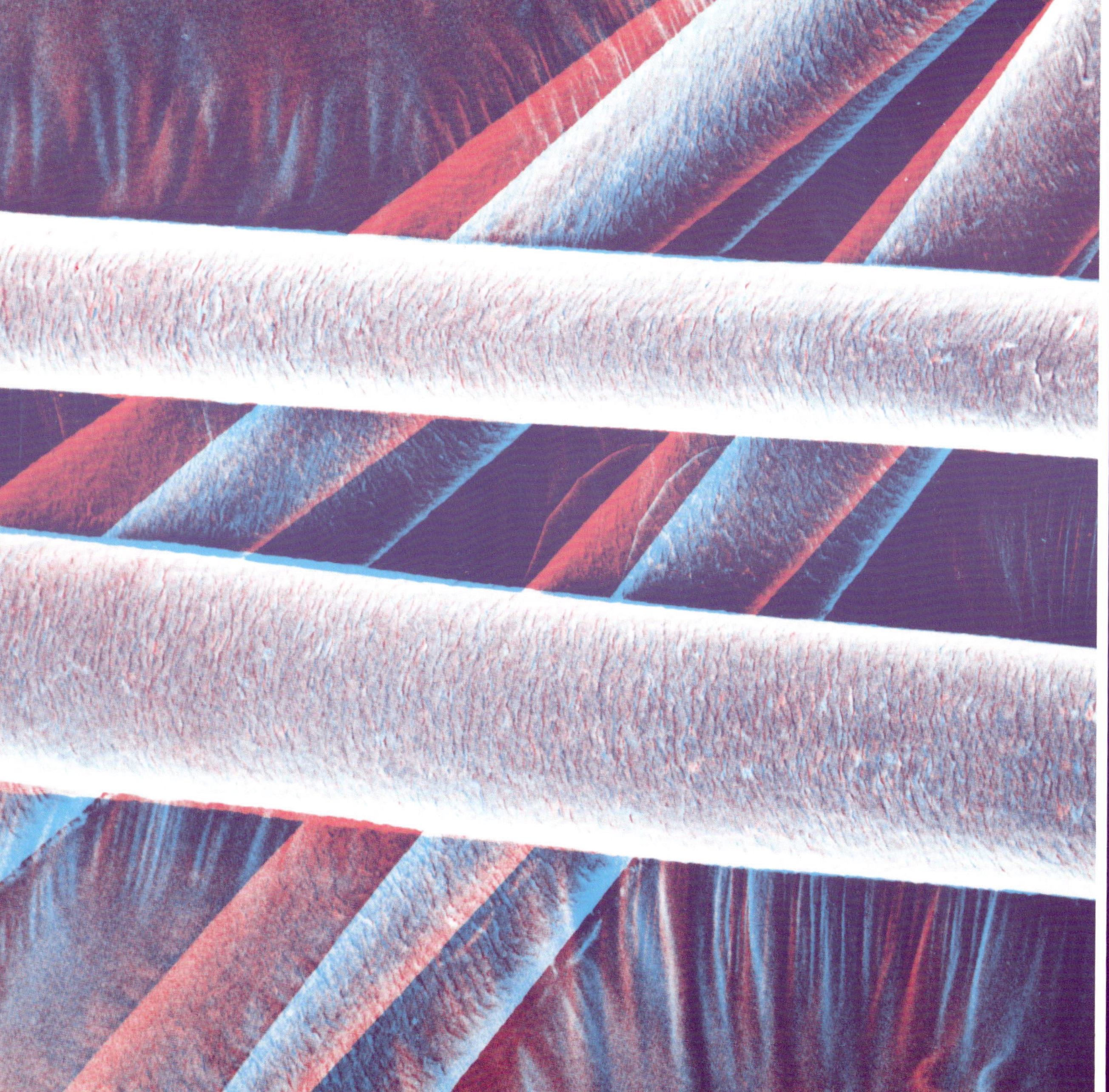

2-D: *150X*↗; 3-D: *300X*→

Plate 3: Nylon Stocking

By the middle of the 16th century, breeches had evolved into separate items of clothing that were knitted much like gloves. The stocking loom was invented in England in the late 16th century and a variety of materials, including silk and cotton, were used in the manufacture of stockings. Nylon was invented in 1938 by DuPont chemist Wallace Carothers and was found to be an ideal material for these clothing articles. The first nylon stockings were placed on sale in New York City on May 15, 1940; within only a few hours, over four million pairs had been sold.

Bild 3: Nylonstrumpf

Etwa in der Mitte des 16. Jahrhunderts hatten sich Hosen und Strümpfe zu eigenständigen Bestandteilen der Kleidung entwickelt, und Strümpfe wurden ähnlich wie Handschuhe gestrickt. Maschinen für die Herstellung von Strümpfen wurden im späten 16. Jahrhundert in England erfunden, und Strümpfe wurden aus vielen unterschiedlichen Materialien gewirkt, wie etwa Seide oder Baumwolle. 1938 entwickelte der Chemiker Wallace Carothers bei der Firma DuPont die Nylon-Faser, die sich schnell als ideal für die Herstellung von Kleiderstoffen erwies. Als am 15. Mai 1940 in New York zum ersten Mal Nylonstrümpfe zum Verkauf angeboten wurden, waren innerhalb weniger Stunden vier Millionen Paar verkauft.

2-D: *12X*↗; 3-D: *24X*→

Plate 4: Iguana Skin

Depending on a reptile's size, its skin may develop small granular scales or larger plate-like scales that either adjoin one another or overlap. Scales may be smooth or ridged and serve primarily to reduce water loss and keep the animal warm. Some of them may also be modified into sharp spines for protection, as shown in this micrograph of a section of tail skin molted by an iguana. You can see here why a predator would be discouraged from attacking this animal.

Bild 4: Leguan-Haut

Abhängig von der Größe eines Reptils kann seine Haut aus kleinen, körnigen oder größeren, plattenförmigen Schuppen bestehen, die entweder dicht aneinanderliegen oder sich überlappen. Die Schuppen können glatt oder schrundig sein, und sie schützen das Tier vor dem Verlust von Feuchtigkeit und Wärme. Schuppen können auch spitz zulaufen und als Stacheln dem Schutz des Tieres dienen. Das Bild zeigt ein Stück Haut vom Schwanz, den ein Leguan bei der Häutung abgestreift hat. Beim Betrachten des Bildes wird verständlich, warum ein Feind den Mut verliert, einen Leguan anzugreifen.

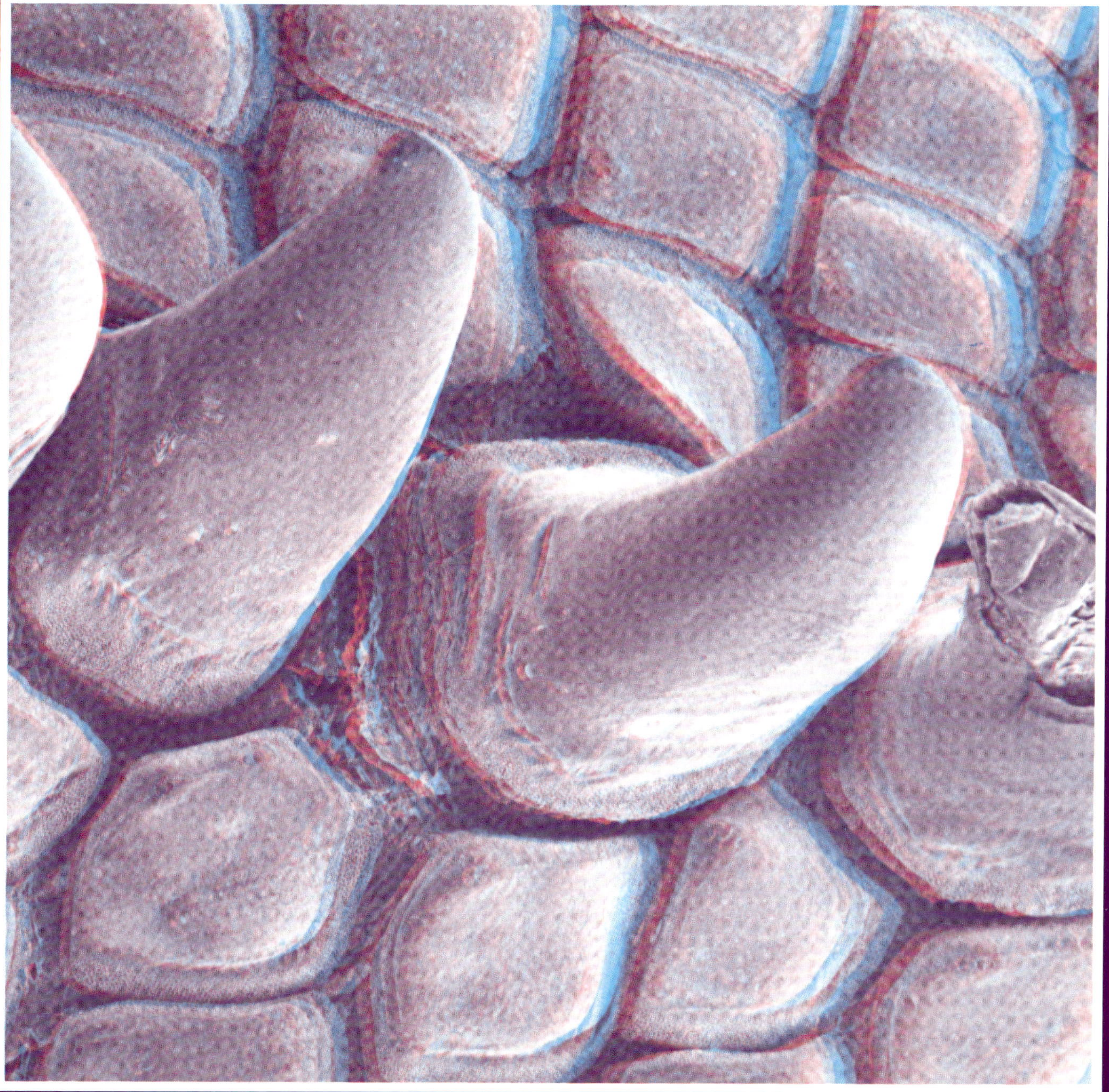

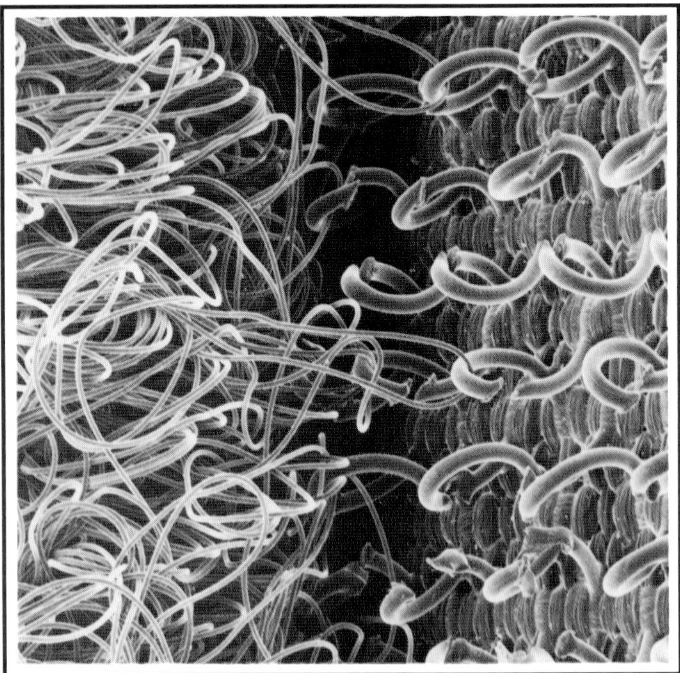

2-D: *15X*↗; 3-D: *30X*→

Plate 5: Velcro

This widely-used fastening device was developed in 1956 by Georges de Mestral of Switzerland after he noticed that the many small hooks on thistles clung tightly to clothing. He duplicated this by developing two nylon strips that stick tightly to each other when pressed together, but that can also be easily pulled apart. He called his invention "Velcro", from the French words *vel*ours (velvet) and *cro*chet (hook); one strip is covered with tiny hooks and the other with tiny loops, as shown in this micrograph. For an example of the natural model upon which this man-made fastener is based, see the Enchanter's Nightshade seed in Plate 22.

Bild 5: Kletterverschluß

Diese vielverwendete Verschlußart wurde 1956 von dem Schweizer Georges de Mestral entwickelt, nachdem er beobachtet hatte, wie sich Disteln mit ihren zahlreichen Widerhäkchen fest an die Kleidung klammern. Er ahmte diesen Effekt nach, indem er zwei Nylonstreifen entwickelte, die fest zusammenhalten, nachdem sie aufeinandergedrückt worden sind, aber doch wieder voneinander getrennt werden können. De Mestral nannte seine Erfindung „Velcro", nach den französichen Wörtern „*vel*ours" (Samt) und „*cro*chet" (Haken). Einer der Streifen ist mit winzigen Häkchen bedeckt, und der andere mit kleinen Schlingen, wie das Bild zeigt. Bild 22 zeigt ein gutes Beispiel aus der Natur, auf dem das Prinzip des Klettverschlusses beruht.

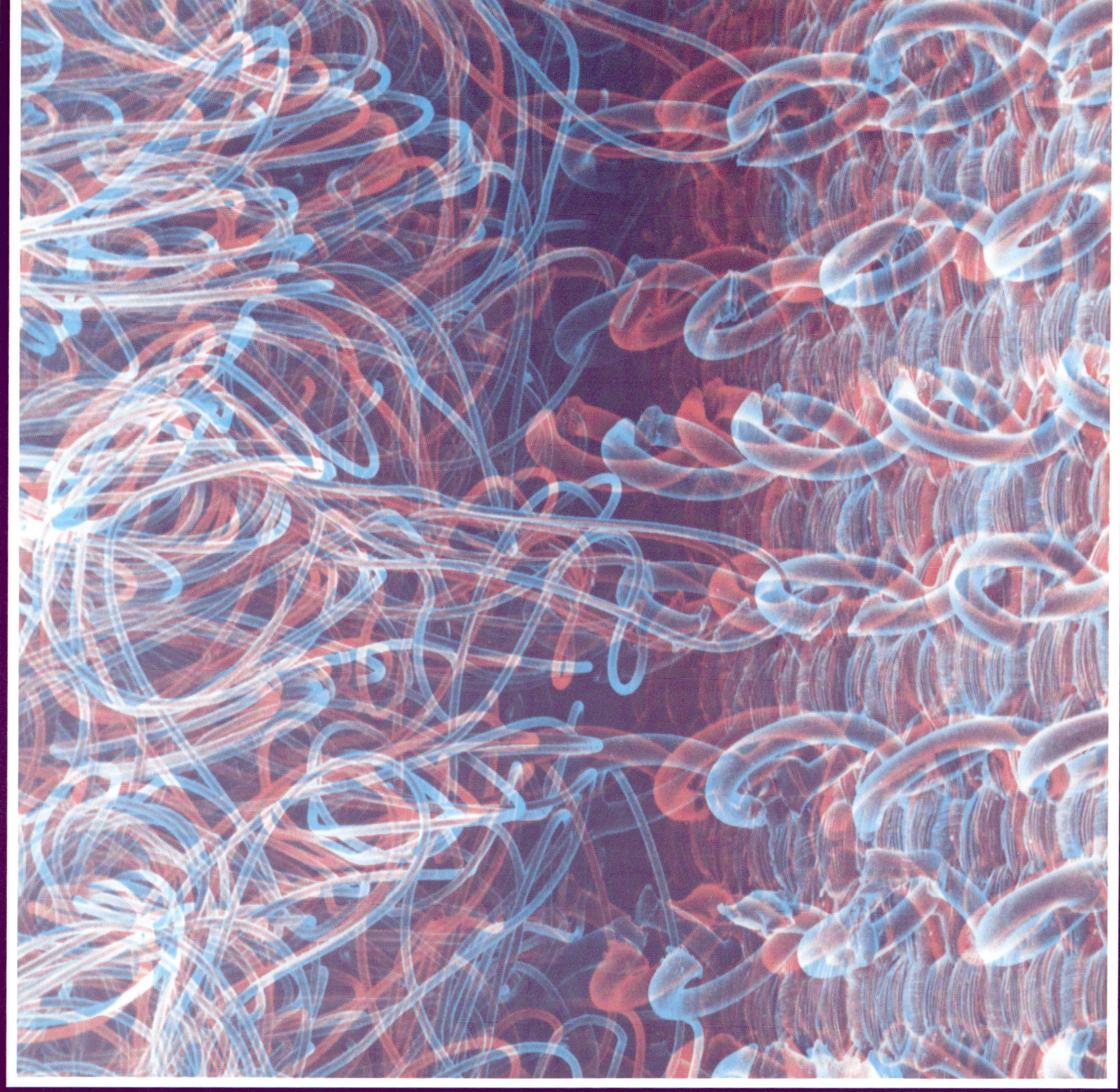

2-D: *650X*↗; 3-D: *1300X*→

Plate 6: Eggshell

This micrograph is one of the more surprising pictures in this collection. We usually think of an eggshell as having almost perfectly smooth inside and outside surfaces, but even a quick glance at this image (which is magnified 1300 times) brings to mind some sort of very rough, multi-layered, net-like structure with sharp spines and protrusions. This is the inside surface of a chicken's eggshell, which is made of calcite, one of the crystalline forms of calcium carbonate. In the lower center, you can see one of the pores through which the developing embryo receives oxygen and releases carbon dioxide during incubation.

Bild 6: Eierschale

Dieses Bild wird beim Betrachter sicher Verblüffung auslösen. Normalerweise glaubt man, eine Eierschale habe innen und außen eine glatte Oberfläche. Aber schon ein kurzer Blick auf das Bild (das eine Vergrößerung von 1300 hat) zeigt einen groben, vielschichtigen und netzförmigen Aufbau mit scharfen Stacheln und Vorsprüngen. Es handelt sich um die innere Oberfläche der Schale eines Kückens. Sie besteht aus Kalzit, einer der kristallinen Formen des Kalziumkarbonats. Unterhalb der Bildmitte erkennt man eine der Poren, durch die der sich entwickelnde Embryo Sauerstoff erhält und durch die während der Bebrütung Kohlendioxid entweicht.

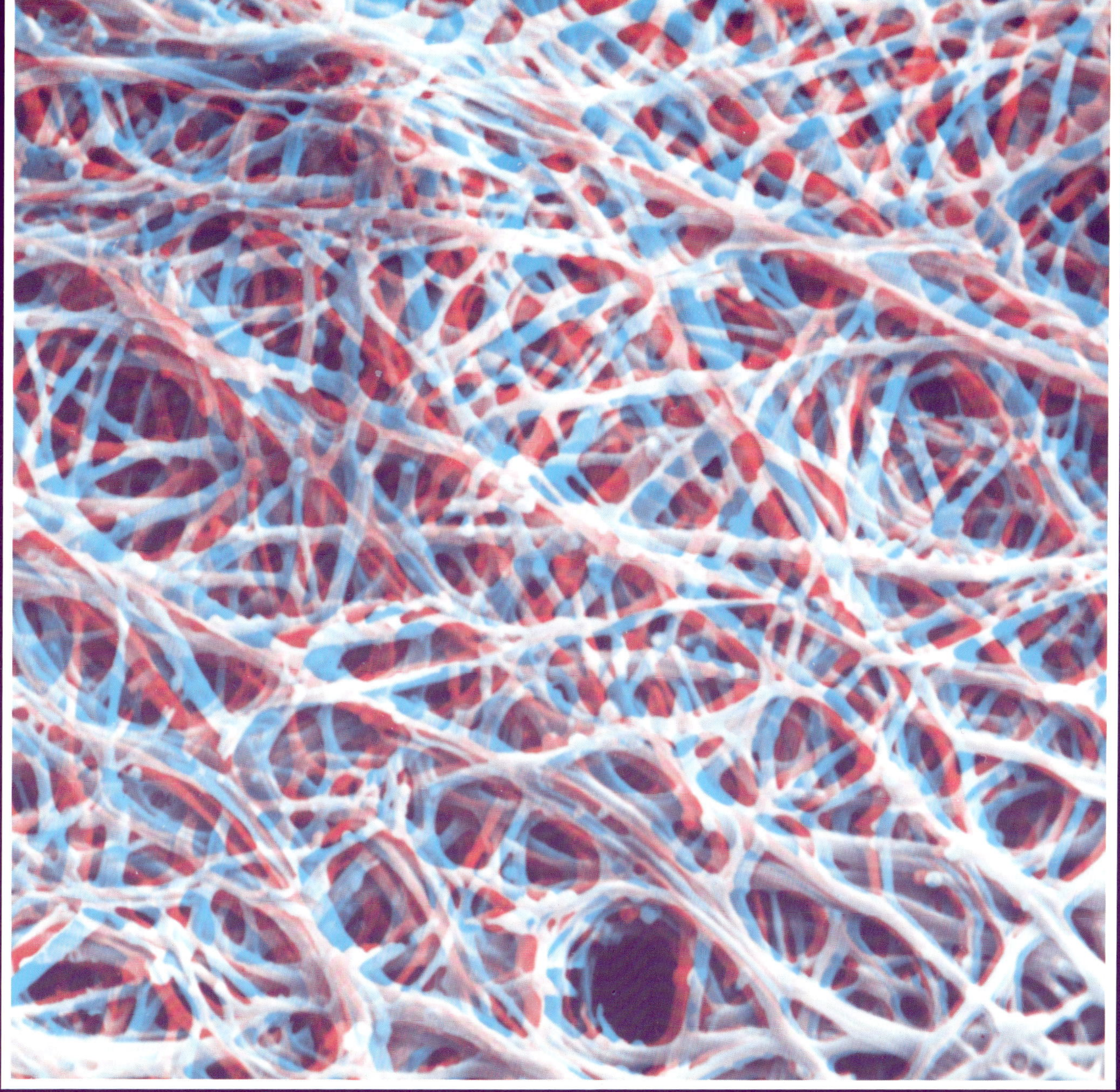

2-D: *105X*↗; 3-D: *210X*→

Plate 7: Goose Feather

Scientists believe a small carnivorous dinosaur was the common ancestor of all birds in the world today. Thus, feathers have evolved from the dinosaur's reptilian scales, becoming elongated and lighter as they adapted to allow flight while still maintaining their original purpose of insulation. Like fingernails, hair, and scales, feathers are made of keratin in the form of a long central shaft, or quill, from which project vanes, barbs, and barbules. The downy feathers of chicks are fluffy for the greatest insulation, but the elongated barbs and barbules of the feathers of adult birds interlock to create an intricate and very efficient aerodynamic structure.

Bild 7: Gänsefeder

Die Wissenschaft hält einen kleinen, fleischfressenden Dinosaurier für den gemeinsamen Vorfahren aller heute existierenden Vögel. Wenn dem so ist, dann haben sich die Federn aus den Hautschuppen dieses Dinosauriers entwickelt. Sie wurden im Verlauf der Entwicklungsgeschichte länger und leichter, um das Fliegen zu ermöglichen, während sie aber immer noch ihre ursprüngliche Schutzfunktion behielten. Ähnlich den Fingernägeln, Haaren und Schuppen, bestehen Federn aus Keratin. Sie haben einen langen, zentralen Schaft, den Kiel, von dem aus die Federstrahlen abzweigen. Die Daunenfedern eines Kückens sind flauschig, weil sie gut isolieren müssen, wogegen sich die langen Strahlen der Federn erwachsener Vögel ineinanderhaken und eine komplizierte und aerodynamisch günstige Struktur bilden.

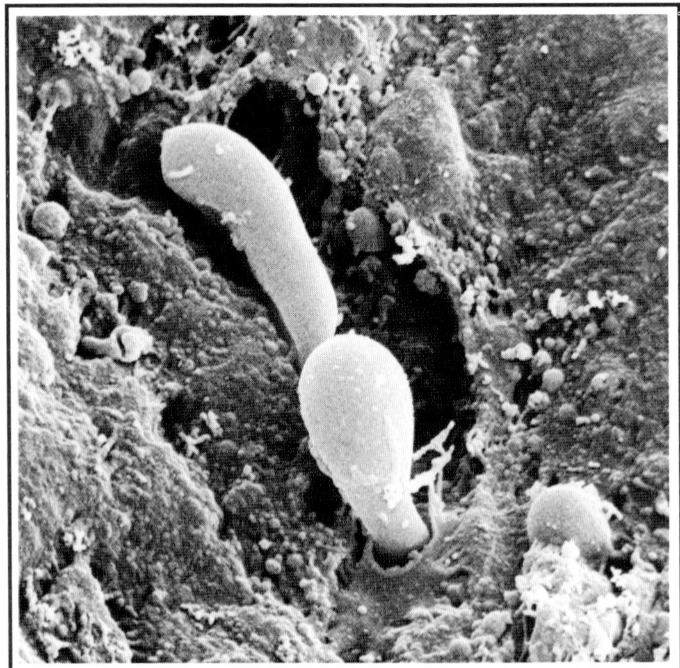

2-D: *5200X*↗; 3-D: *10400X*→

Plate 8: Thyroid Hormones

The thyroid is an important gland that secretes the hormones that regulate metabolism (which affects every cell in the body) and contribute to the development of the brain and bones in children. Shown here is the first stage of the two-stage hormone production process: thyroid protein is being secreted into a storage chamber, called a follicle, from the cells lining the chamber's walls. The second stage of the process is completed in the cell walls after the protein has been stored in the follicle, and it is only then that the hormone is released into the body.

Bild 8: Schilddrüsenhormone

Die Hormone der Schilddrüse regeln den Stoffwechsel, dem jede Zelle des Körpers unterliegt, und sie sind wichtig für die Entwicklung des Gehirns und des Knochenbaus bei Kindern. Das Bild zeigt die erste Stufe des zweistufigen Entstehungsprozesses eines Hormons: Die Schilddrüse sekretiert eine Eiweißverbindung aus den Zellen, die die Wände einer Vorratskammer (Follikel) auskleiden. Die zweite Stufe des Vorganges läuft in den Zellwänden ab, nachdem das Eiweiß in dem Vorratsgefaß gesammelt wurde, und erst dann wird das Hormon in die Blutbahn abgegeben.

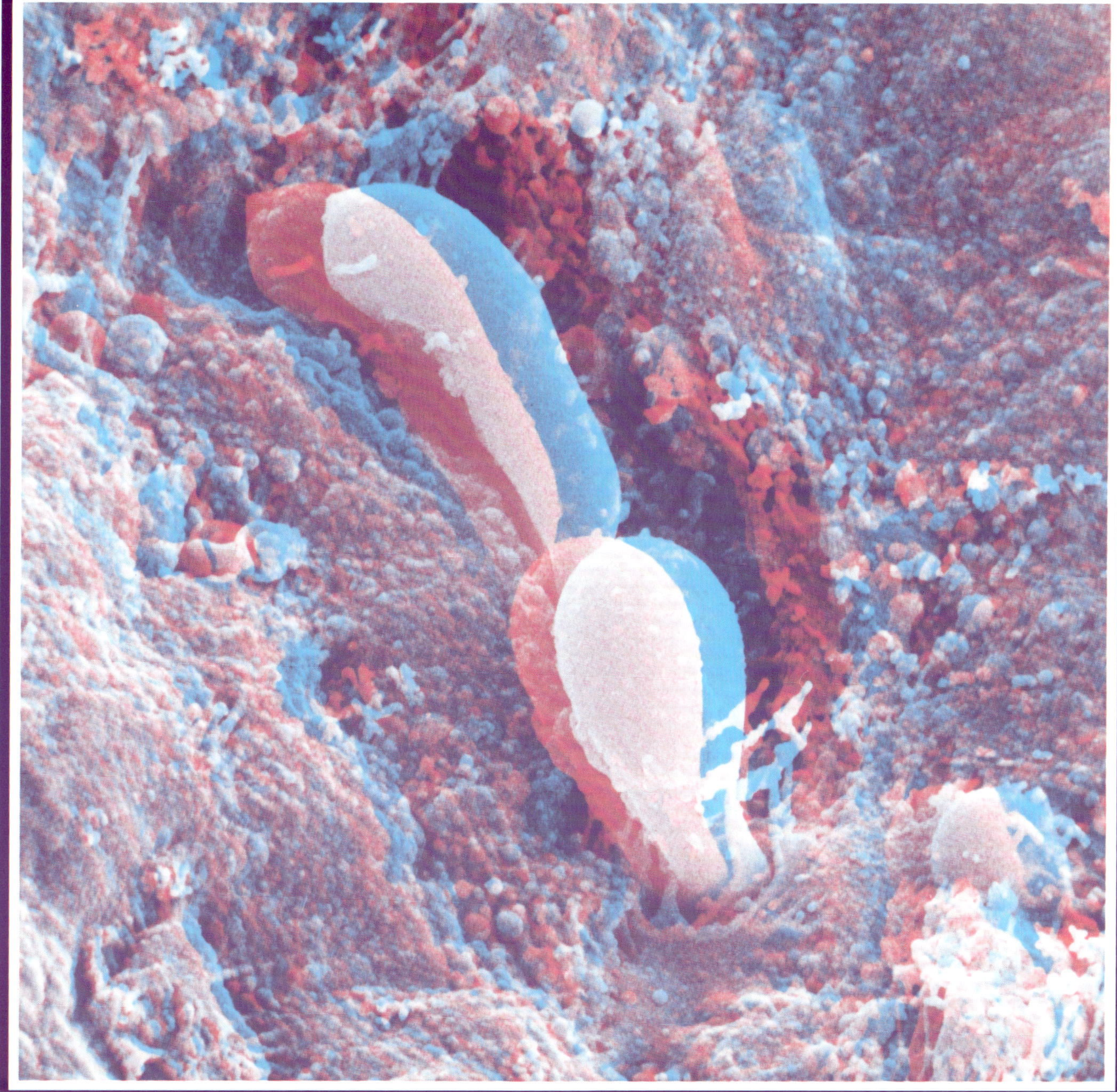

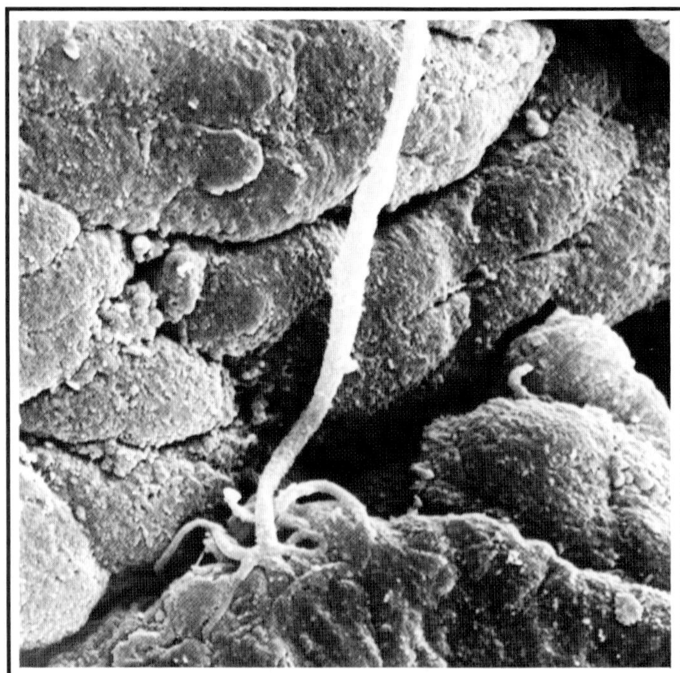

2-D: *3100X*↗; 3-D: *6200X*→

Plate 9: Capillary Network

Capillaries are the smallest blood vessels in the body; some are so small that only single blood cells can flow through them in a line. This micrograph shows the end of a small blood vessel connected to a capillary network in human thyroid tissue. Each hormone storage chamber is surrounded by a network of capillaries that delivers amino acids, iodine, and other substances to the cells of the chamber walls and transports the hormones produced there to the body's circulatory system.

Bild 9: Blutgefäße

Kapillaren sind die kleinsten Blutgefäße im Körper. Manche Kapillaren sind so dünn, daß die Blutkörperchen nur hintereinander hindurchfließen können. Das Bild zeigt das Ende eines kleinen Blutgefäßes, das mit einem Kapillarennetzwerk im Gewebe einer menschlichen Schilddrüse verbunden ist. Jede Hormonspeicherkammer (siehe Bild 8) ist von einem Kapillarennetzwerk umgeben, das die Zellen der Kammerwand mit Aminosäuren, Iod und anderen Stoffen versorgt und die produzierten Hormone in das Kreislaufsystem des Menschen transportiert.

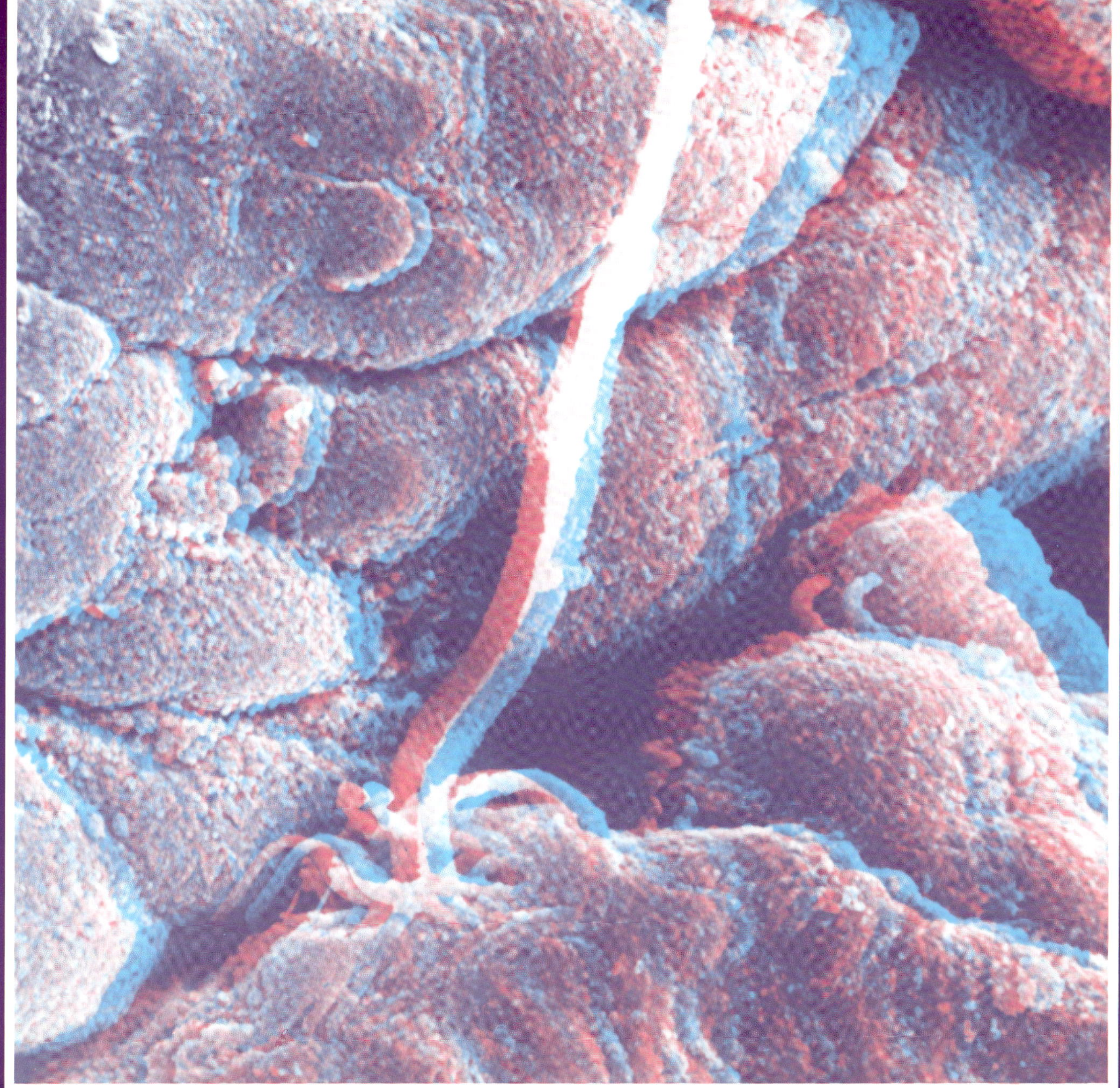

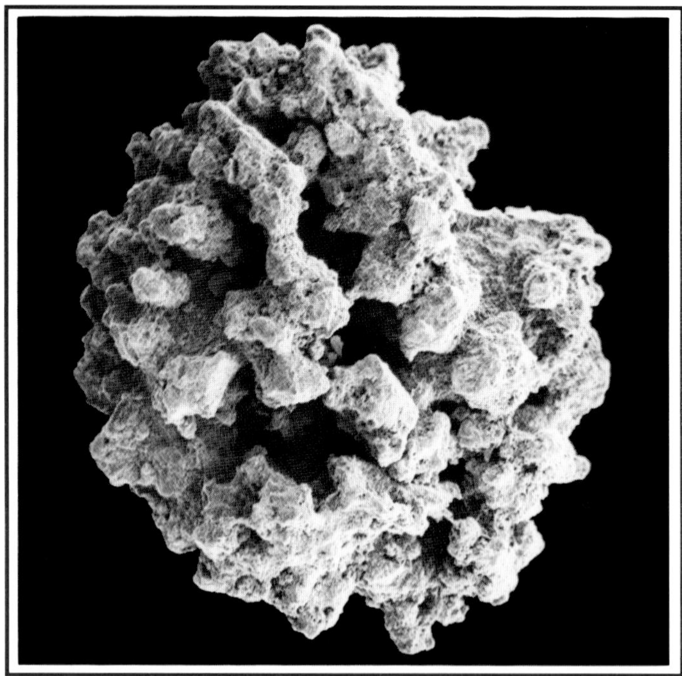

2-D: *25X*↗; 3-D: *50X*→

Plate 10: Pancreatic Stone

The pancreas is also a hormone-secreting gland. It's part of the digestive system, which includes the stomach, intestines, liver, and gall bladder, the latter two of which can also produce similar stones. Pancreatic stones are made primarily of calcium carbonate, proteins, and sugars, and have a knobby white or dull brown appearance. They have been studied since the 17th century and are usually associated with diabetes, chronic alcoholism, tropical pancreatitis and other illnesses. The association of these stones with diabetes may have guided F.G. Banting and C.H. Best to the discovery of insulin in 1921.

Bild 10: Stein der Bauchspeicheldrüse

Auch die Bauchspeicheldrüse sekretiert Hormone. Sie ist Teil des Verdauungssystems, das außerdem Magen, Darm, Leber und Gallenblase umfaßt. Leber und Gallenblase können ähnliche Steine produzieren. Steine der Bauchspeicheldrüse bestehen hauptsächlich aus Kalziumkarbonat, Eiweißen und Zucker und haben ein knubbeliges, weißes oder mattbraunes Aussehen. Diese Steine werden schon seit dem 17. Jahrhundert erforscht. Sie treten in der Regel im Zusammenhang mit Zuckerkrankheit, chronischem Alkoholismus, Entzündung der Bauchspeicheldrüse und anderen Krankheiten auf. Der Zusammenhang zwischen diesen Steinen und Zuckerkrankheit könnte F.G. Banting und C.H. Best im Jahre 1921 zur Entdeckung des Insulins geführt haben.

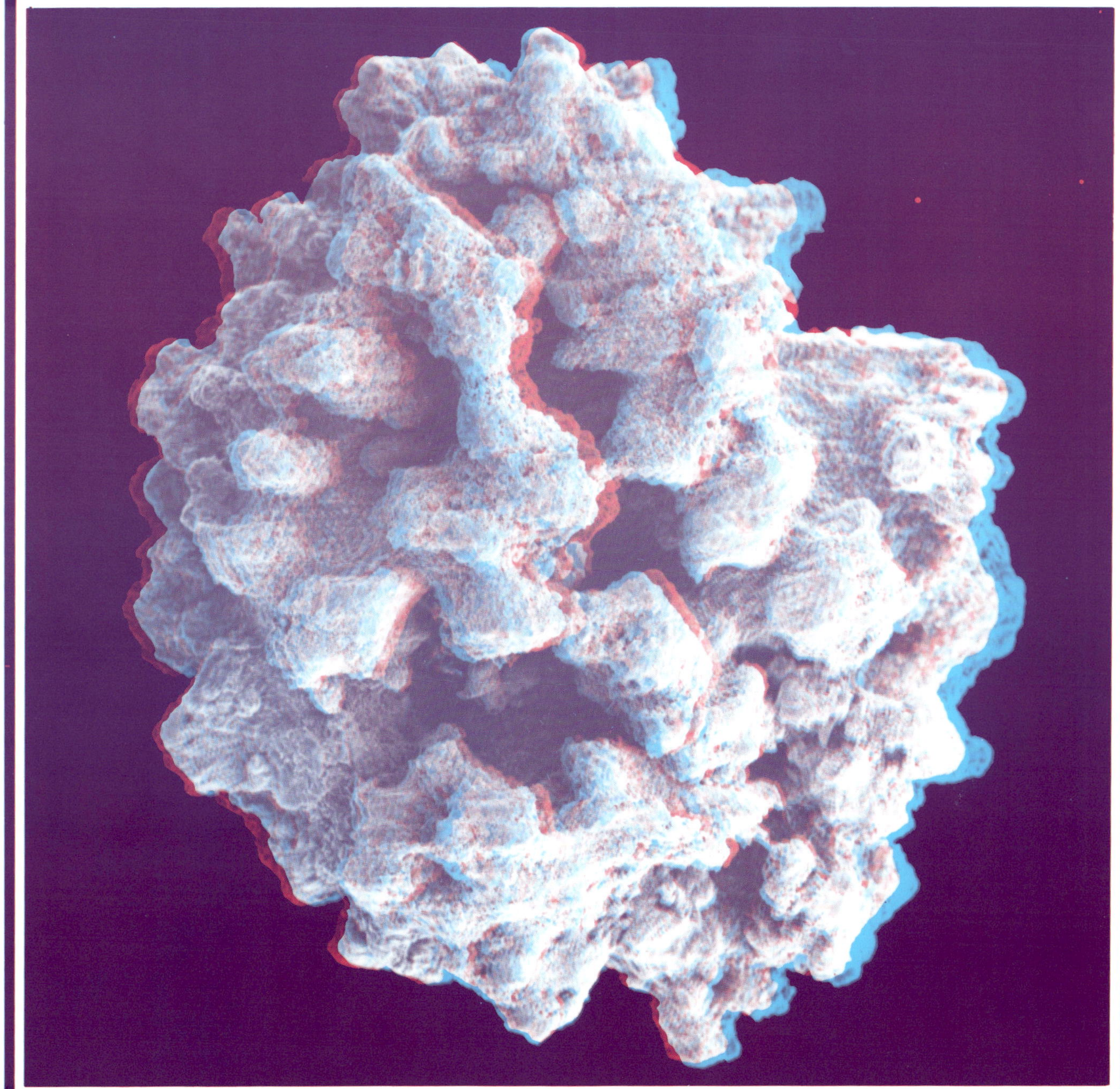

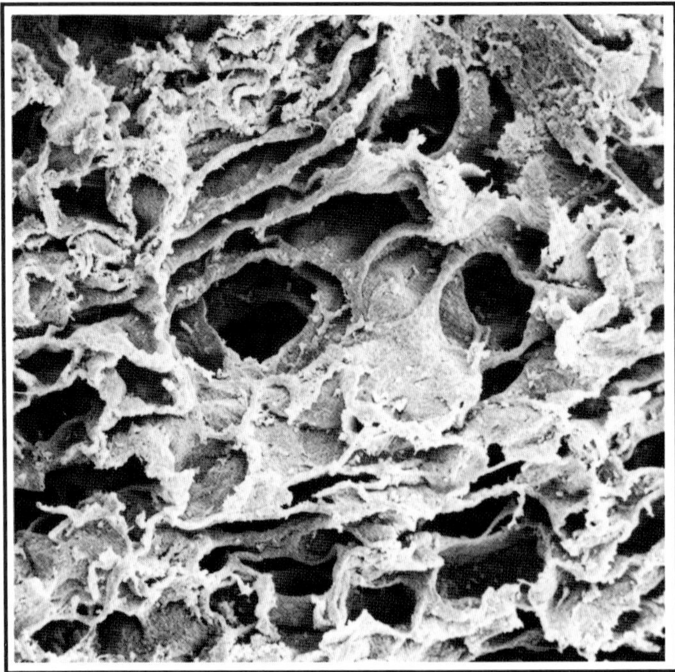

2-D: *75X*➚; 3-D: *150X*➔

Plate 11: Lung Tissue

The spongy appearance of this human lung tissue is due to the many air sacs, or alveoli, separated from each other by extremely thin membranes. A system of branching tubes (the bronchi and bronchioles) conducts air to the alveoli, where oxygen and carbon dioxide are exchanged between the alveoli and blood in the surrounding capillary networks. The pulmonary arteries and arterioles that conduct the blood to the lungs have many elastic fibers arranged in a circular pattern to permit expansion and contraction during breathing.

Bild 11: Lungengewebe

Das schwammähnliche Aussehen dieses Gewebes der menschlichen Lunge rührt von den zahlreichen Luftkammen, oder Alveolen, die durch äußerst dünne Membranen voneinander getrennt sind. Ein System sich verzweigender Röhren (die Bronchien und Branchiolen) leitet die Luft zu die Alveolen, wo Sauerstoff und Kohlendioxid zwischen diesen Alveolen und dem Blut in den benachbarten Kapillarnetzen ausgetauscht werden. Die Lungenarterien und -arteriolen, die das Blut zu den Lungen führen, haben viele elastische Fasern, die ringförmig angeordnet sind und das Ausdehnen und Zusammenziehen beim Atmen ermöglichen.

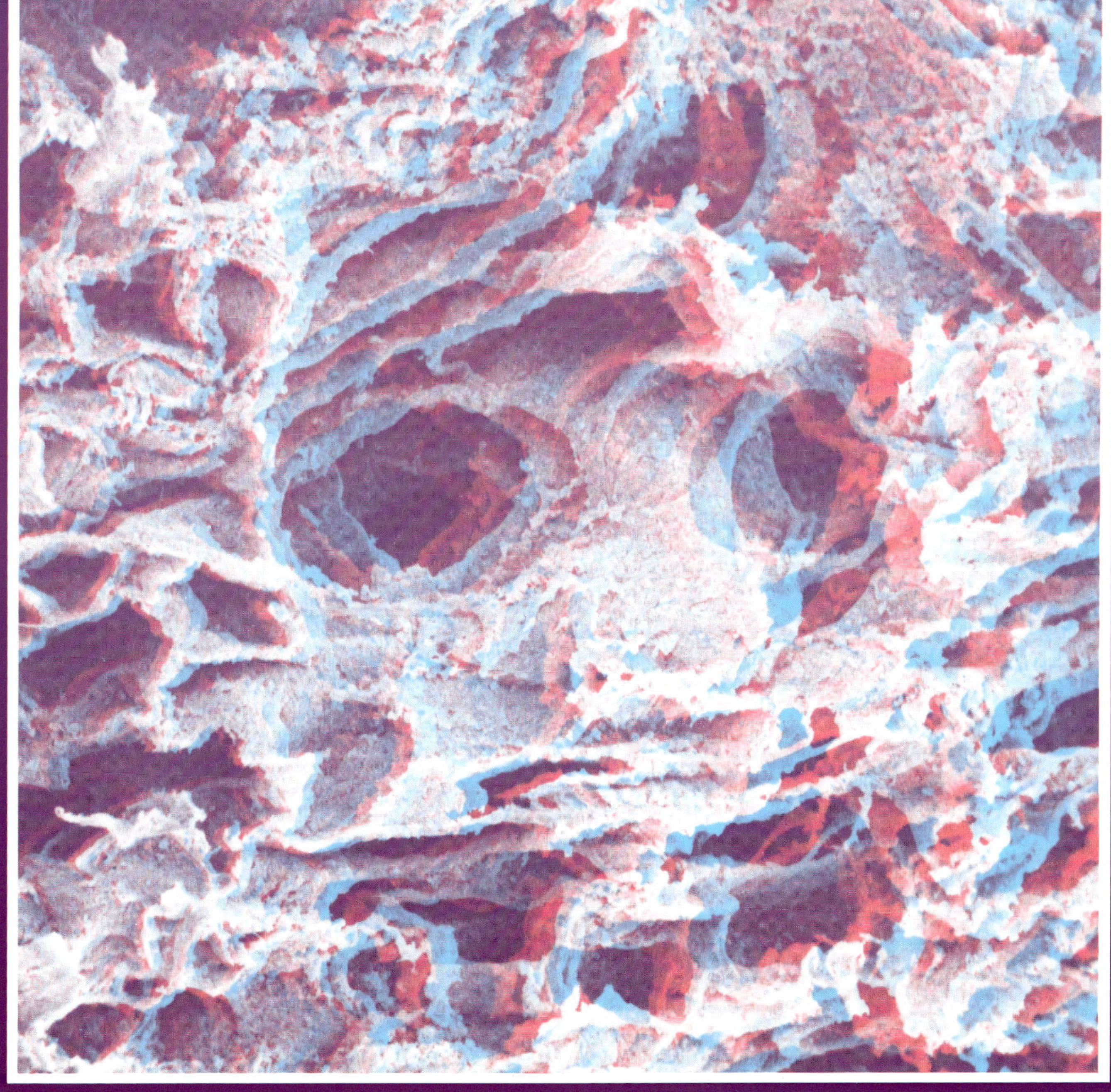

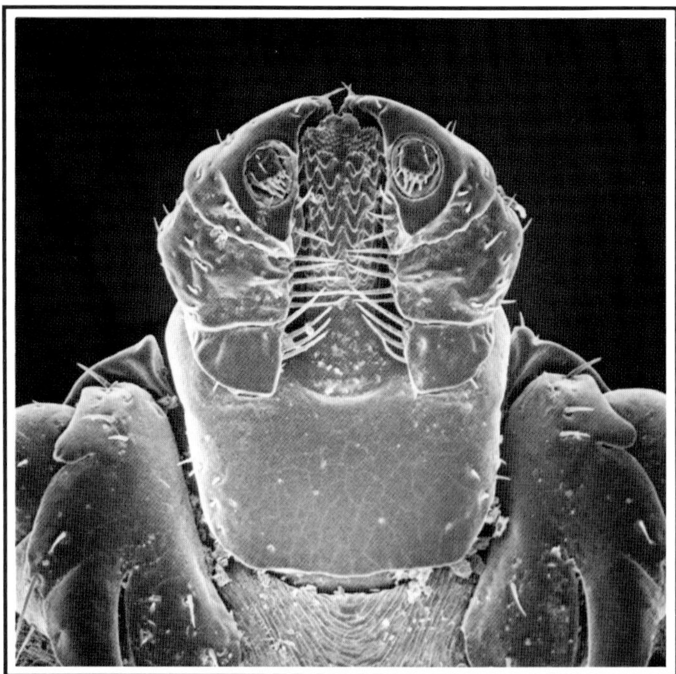

2-D: *70X*↗; 3-D: *140X*→

Plate 12: Tick Portrait

Ticks are parasitic insects that live on the blood of many land animals, and they can transmit serious illnesses, such as Lyme Disease and Rocky Mountain Spotted Fever, to humans. Ticks climb to the tops of bushes to lie in wait for their victims, which they detect by humidity and smell through special sensory organs that are quite similar to the whiskers of the barnacle shown in Plate 1. When an animal passes by, the ticks jump on to have a feast of blood, after which they fall to the ground to molt or lay their eggs.

Bild 12: Porträt einer Zecke

Zecken sind schädliche Insekten, die vom Blut vieler Landtiere leben. Sie können gefährliche Krankheiten auf Menschen übertragen, wie z.B. Lyme-Borreliose, eine tückische Krankheit mit einer ganzen Bandbreite von Beschwerdebildern, oder Fleckfieber (Felsengebirgsfieber). Zecken warten oben in Büschen auf ihre Opfer, die sie an der Feuchtigkeit und am Geruchs mittels spezieller Sinnesorgane erkennen. Diese Organe ähneln stark den Härchen der Rankenfüßer (Bild 1). Wenn sich ein Tier nähert, lassen sie sich auf dieses fallen und saugen sich mit seinem Blut voll. Danach fallen sie zu Boden, wo sie sich häuten oder ihre Eier ablegen.

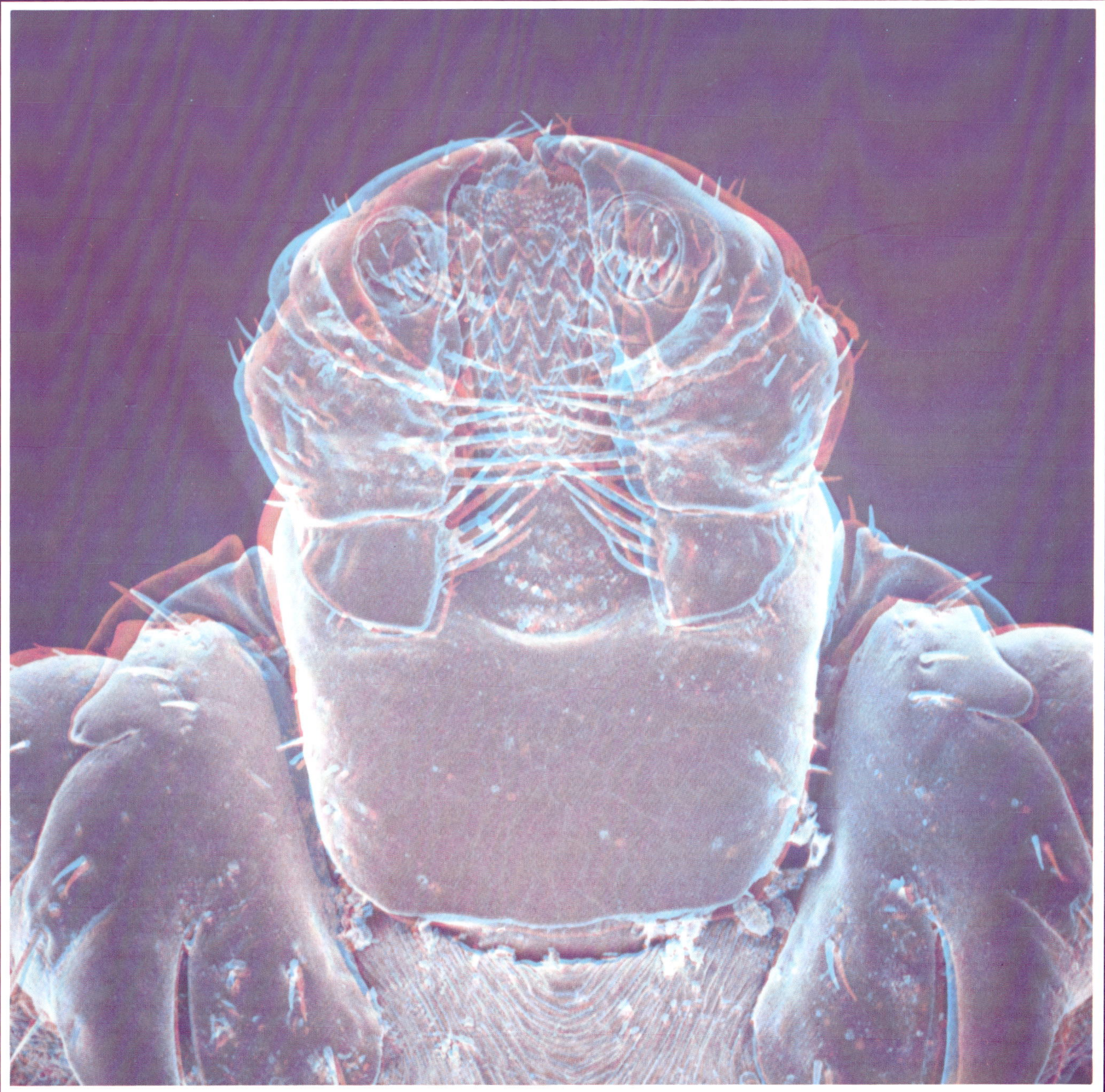

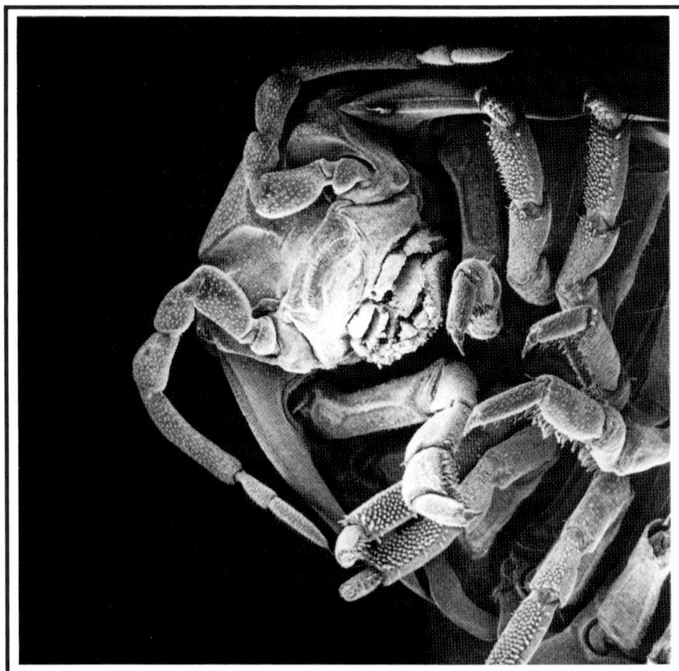

2-D: *12X*➚; 3-D: *24X*→

Plate 13: Millipede Portrait

There are about 800 species of millipedes, which have from nine to 200 leg-bearing segments and no wings; although they lack the true compound eyes of other arthropods, they have excellent antennae for the senses of smell and taste which they use to find food as they burrow through soil and leaf litter. The millipede shown here may look ferocious, but like most others of its kind, it's an herbivore that lives on plants and organic debris. You can clearly see its well-developed antennae and the claws on its legs, but its small mouth parts and insignificant eyes are less apparent.

Bild 13: Porträt eines Tausendfüßers

Es gibt ungefähr 800 Arten Tausendfüßer, die zwischen neun und 200 beintragende Körpersegmente und keine Flügel haben. Sie verfügen auch nicht über die Art Sehorgane anderer Gliederfüßer, dafür besitzen sie aber ausgezeichnete Organe für den Geruch und den Geschmack, mit deren Hilfe sie ihre Nahrung finden, während sie Erdreich und altes Laub durchwühlen. Der Tausendfüßer auf dem Bild mag zwar furchterregend aussehen, aber wie die meisten seiner Art ist er ein Pflanzenfresser. Seine gut entwickelten Antennen und die Klauen an den Füßen sind gut zu erkennen, weniger gut zu sehen sind das kleine Maul und die kaum entwickelten Sehorgane.

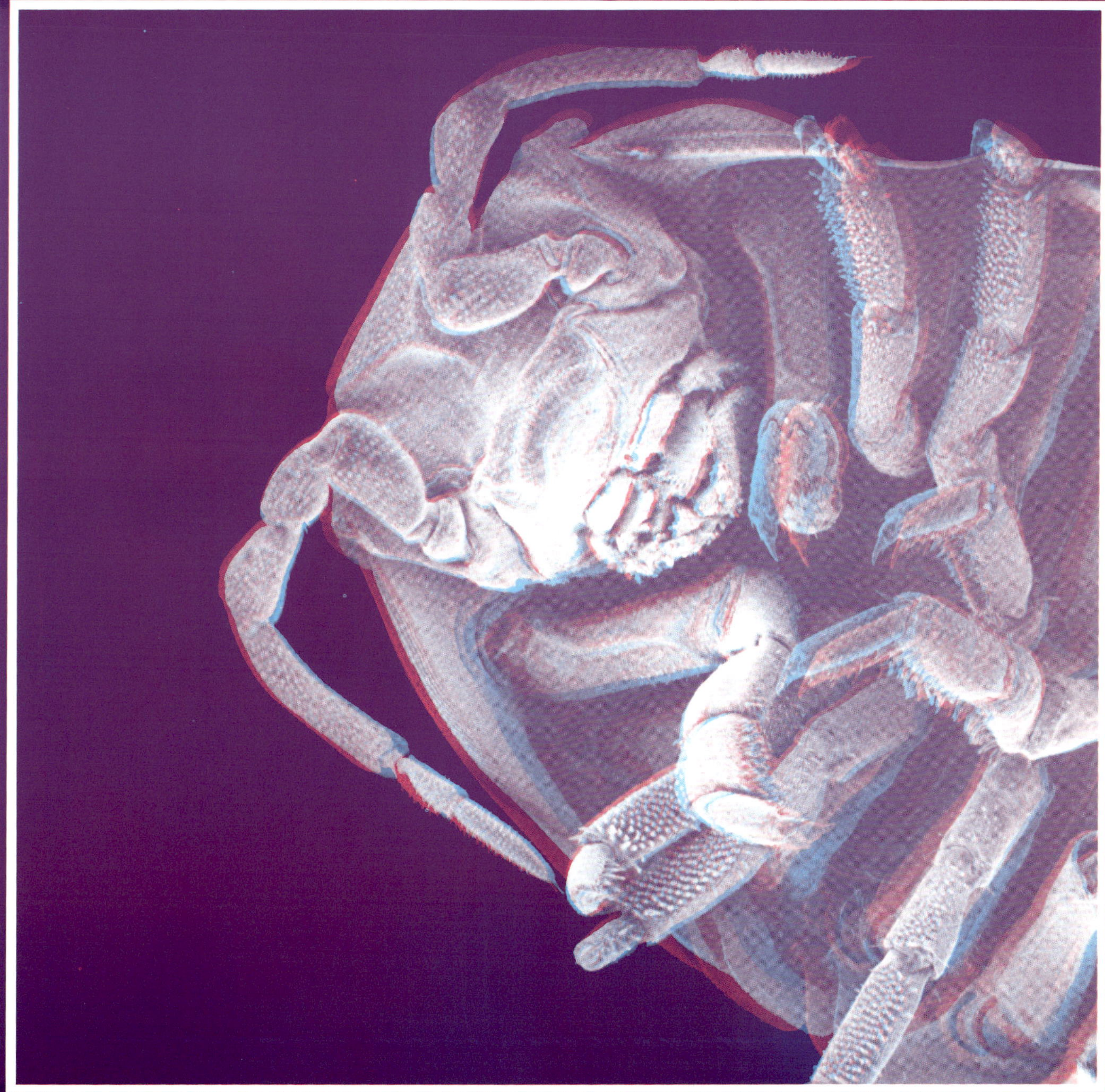

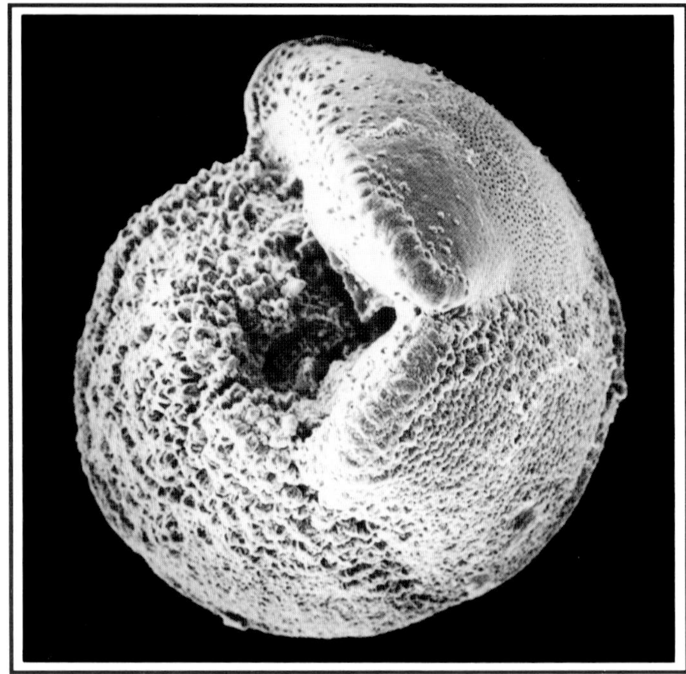

2-D: *100X*↗; 3-D: *200X*→

Plate 14: Planktonic Foraminifer

Foraminifera are single-celled marine protozoans that have a calcium carbonate shell. The fossils of larger species in the limestone of the Egyptian pyramids were once mistaken for lentils eaten by the pyramid builders. Information about ancient oceans and climates provided by studying these tiny shells helps scientists understand the oceans of the last 100 million years. The microscopic specimen shown in this image was collected in a plankton tow during an oceanographic expedition near the Mid-Atlantic Ridge, which is part of a system of submerged mid-ocean ridges and rifts that encircles the earth.

Bild 14: Wurzelfüßer (Foraminifere)

Foraminiferen sind einzellige Tiere (Protozoen) des Meeres mit einer Kalziumkarbonatschale. Versteinerungen von größeren Arten dieser Einzeller, die man im Kalkstein ägyptischer Pyramiden fand, hielt man eine Zeitlang irrtümlich für Speiselinsen, von denen sich die Erbauer der Pyramiden ernährt haben sollen. Aus Untersuchungen dieser winzigen Schalen können die Wissenschaftler wertvolle Informationen erhalten über die Ozeane und das Klima der letzten 100 Millionen Jahre. Das mikroskopisch kleine Exemplar auf der Abbildung fand man im Plankton während einer ozeanographischen Expedition nahe dem Zentralatlantischen Graben, der Bestandteil eines Systems von ozeanischen Gräben ist, die rund um die Erde verlaufen.

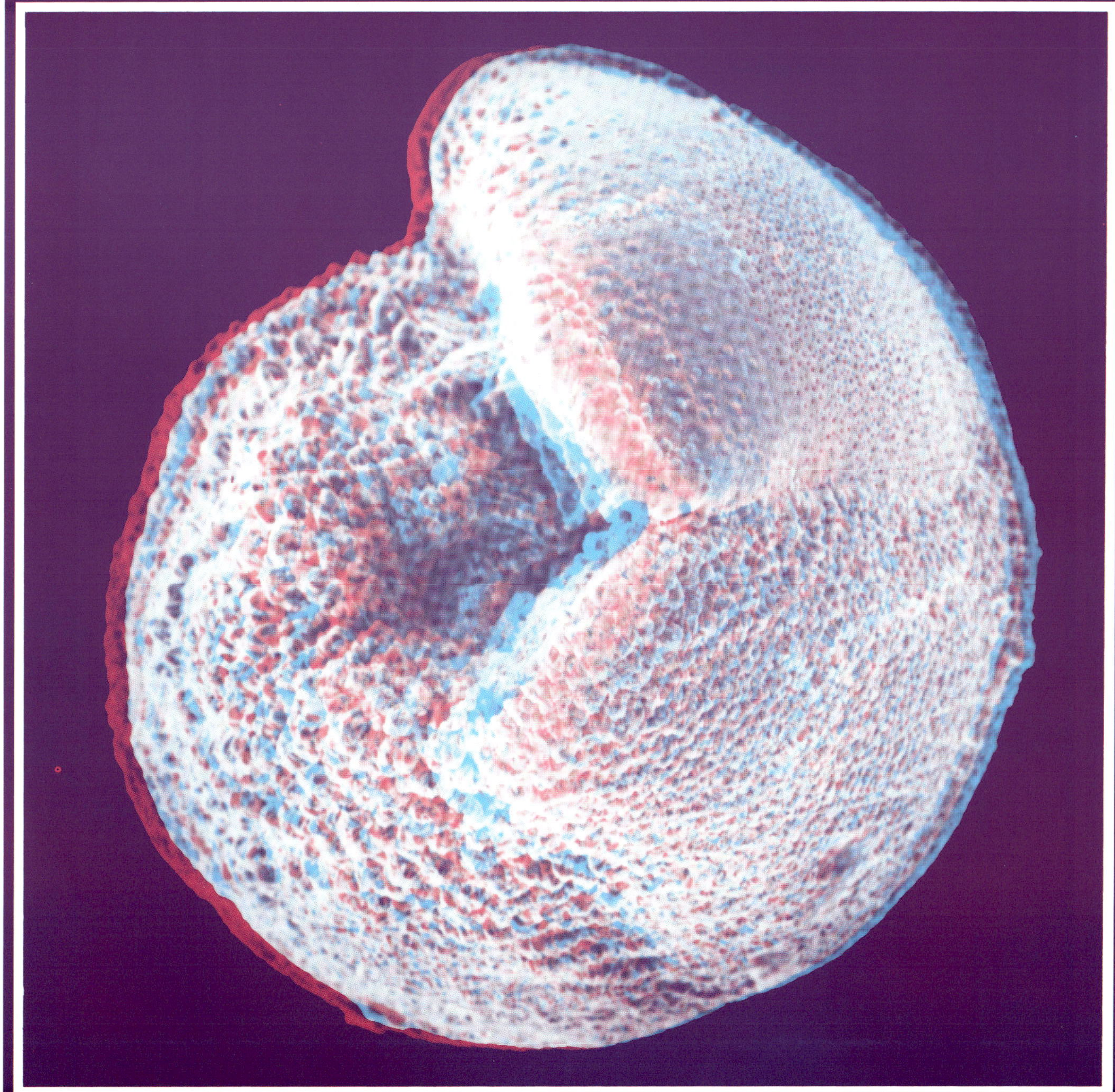

2-D: *200X*↗; 3-D: *400X*→

Plate 15: Fossil Radiolaria

Radiolaria are one-celled marine protozoa that live in all the oceans of the world. The most remarkable feature of these microscopic organisms is their glassy skeletons, which are secreted as ornate shells that surround the central bodies. Each species produces its own uniquely-shaped shell, and many of them are quite elaborate, as you can see in this micrograph of two fossil radiolaria found in a core sample taken from sediments beneath the Ross Sea off the coast of Victoria Land in Antarctica.

Bild 15: Versteinertes Strahlentierchen

Strahlentierchen (Radiolarien) sind Einzeller, die in allen Weltmeeren leben. Besonders bemerkenswert an diesen mikroskopisch kleinen Organismen ist ihr glasiges Skelett, das von den Tierchen ausgeschieden wird und den inneren Körper als kunstvolle Schale umgibt. Jede Art dieser Strahlentierchen erzeugt eine Schale mit einem ganz individuellen Muster. Eindrucksvolle Beispiele hierfür zeigt das Bild zweier versteinerter Strahlentierchen, die in Ablagerungen im Rossmeer vor der Küste Victorialands in der Antarktis gefunden wurden.

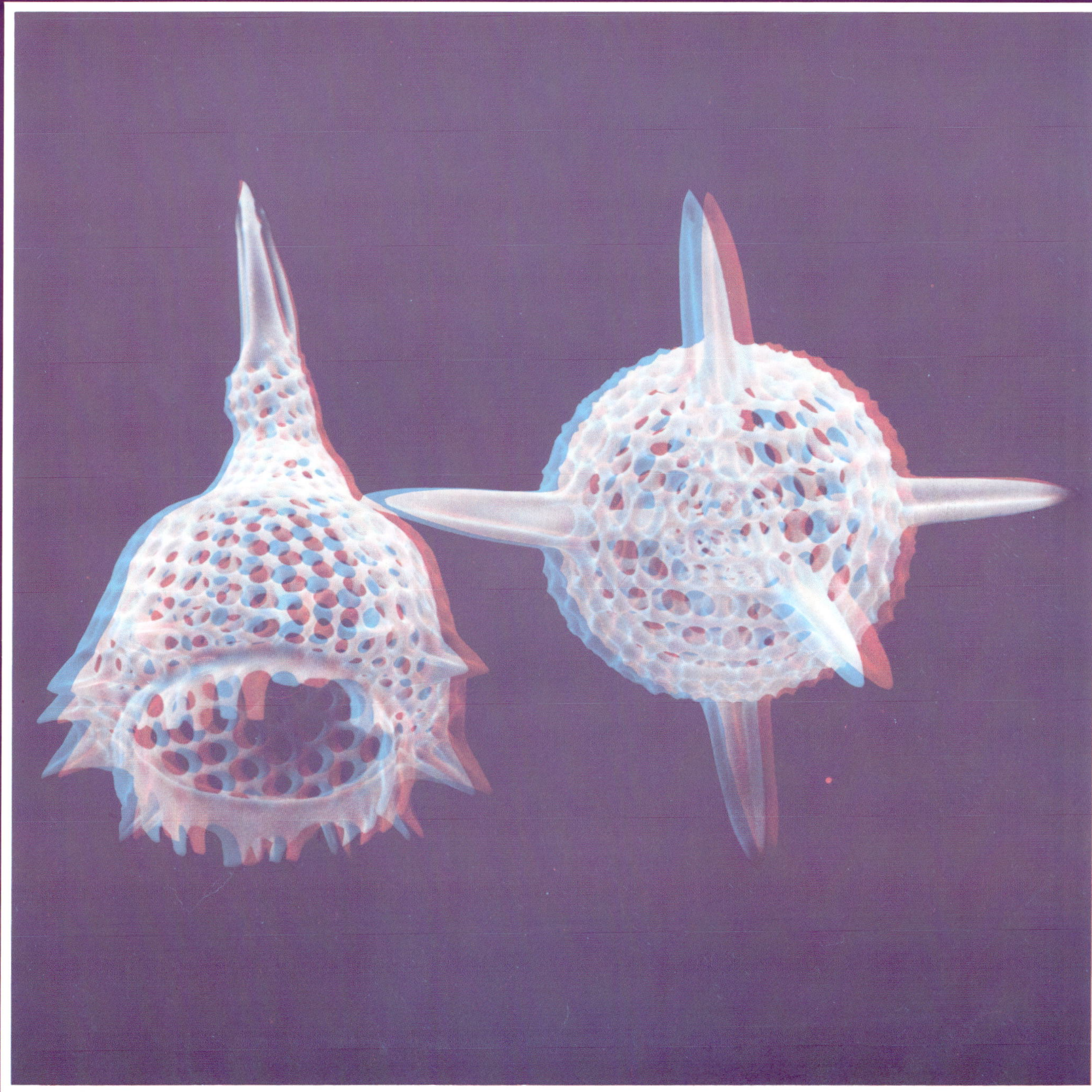

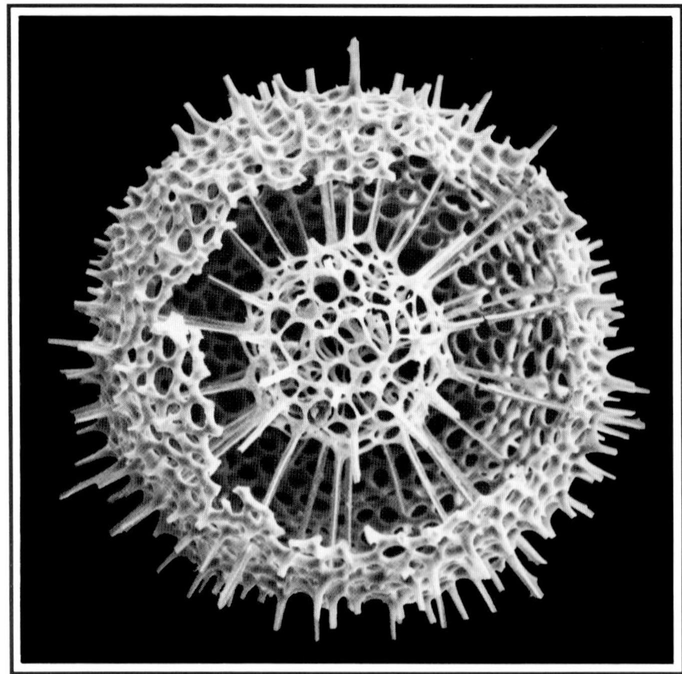

2-D: *350X*↗; 3-D: *700X*→

Plate 16: Fossil Radiolarian

Radiolaria are microorganisms that live at all depths of the oceans and extend far back into geologic time, thus providing us with a very detailed evolutionary record. Their remarkable beauty reminds us that even the simplest life forms can still be extremely complex. This little Antarctic radiolarian fossil – only one one-hundredth inch in diameter – is shown with part of its outer shell broken away, which allows you to see its construction of skeletal spheres, one nested within another. At first glance, there appear to be only two spheres, but if you look very carefully, you'll see even another tiny skeletal sphere inside the smaller sphere.

Bild 16: Versteinertes Strahlentierchen

Strahlentierchen (Radiolarien) sind Mikroorganismen und leben in allen Tiefen der Ozeane. Ihr Auftreten reicht in der Erdgeschichte weit zurück. Daher vermitteln sie uns detaillierte entwicklungsgeschichtliche Informationen. Ihre bemerkenswerte Schönheit ist ein Hinweis darauf, daß selbst einfachste Lebensformen äußerst kompliziert sein können. Bei dem kleinen, antarktischen Strahlentierchen auf dem Bild – sein Durchmesser beträgt nur ein Viertel Millimeter – fehlt ein Teil der äußeren Schale, so daß man die ineinander verschachtelten, kugelschalenartig aufgebauten Skeletteile gut erkennen kann. Auf den ersten Blick sieht man nur zwei Kugelschalen, aber bei näherem Hinsehen erkennt man eine weitere, winzige Kugel im Innerin der kleineren Schale.

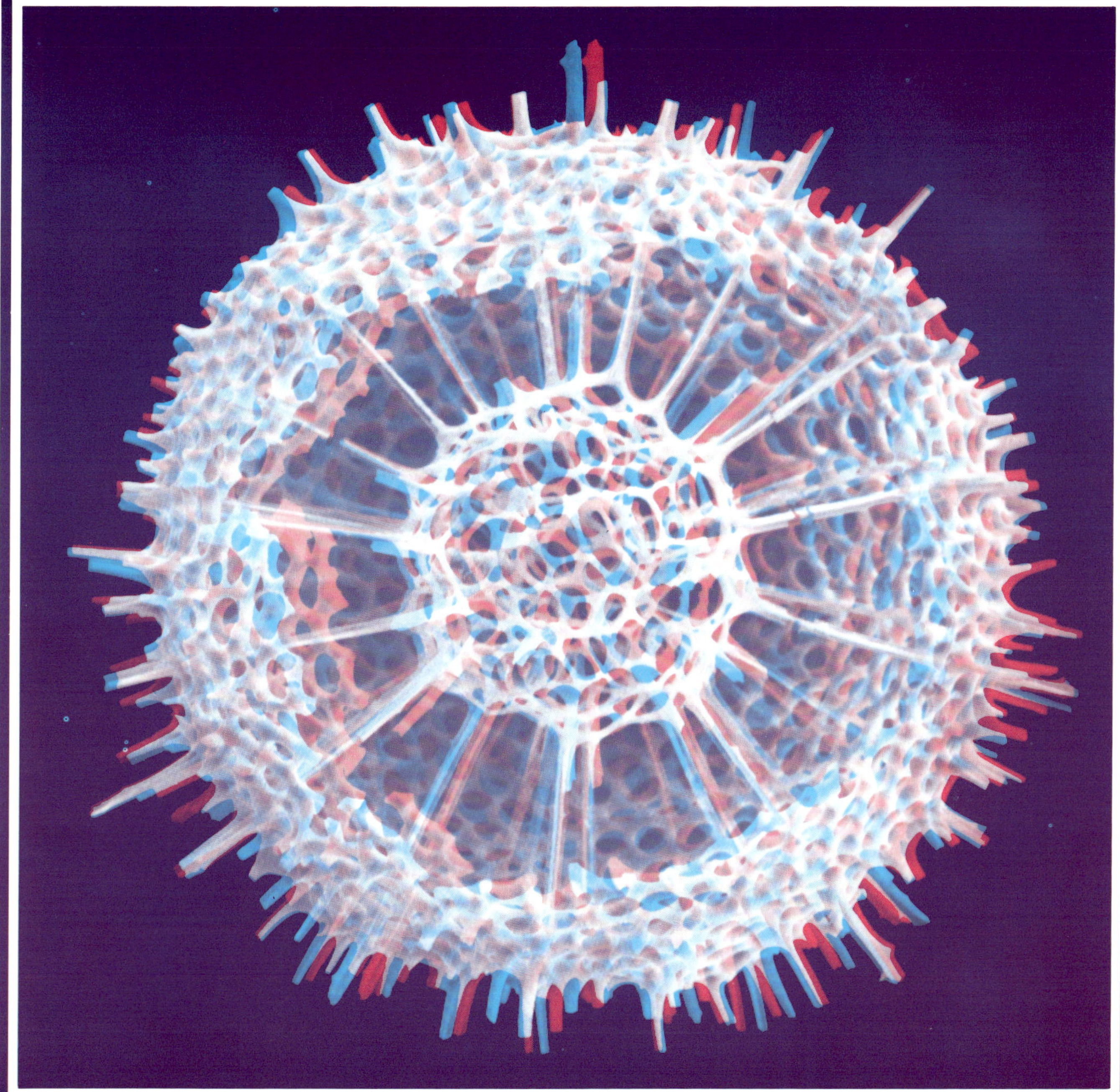

2-D: *3400X*↗; 3-D: *6800X*→

Plate 17: Coccolithophorids

These ultramicroscopic organisms are one-celled algae that live in the subpolar through tropical surface waters of the world's oceans. They are a major source of food for marine animals, as well as atmospheric oxygen. The living cells are covered with small elaborate calcium carbonate plates, called coccoliths, that overlap to protect the cell inside. After the living cell dies, the individual coccoliths separate and settle to the ocean floor, where they form a fine-grained lime mud. Shown here is a cluster of coccolithophorids that was filtered from tropical sea water during an oceanographic expedition.

Bild 17: Kalkalgen

Diese unter dem Lichtmikroskop nicht mehr erkennbaren Organismen sind einzellige Algen und leben in den obersten Schichten der subpolaren bis tropischen Bereiche der Weltmeere. Sie stellen eine wichtige Nahrungsquelle für Meerestiere als auch einen Sauerstoffspeicher dar. Die lebenden Zellen sind mit kleinen, kunstvoll geformten Kalziumkarbonat-Plättchen bedeckt, sog. Kokkolithen, die sich gegenseitig überlappen und so die Zelle im Innerin schützen. Wenn die Zelle abstirbt, trennen sich die Kokkolithen von ihr und sinken auf den Boden des Ozeans, wo sie einen feinkörnigen Kalkschlamm bilden. Das Bild zeigt eine Gruppe von Kalkalgen, die aus tropischem Meereswasser herausgefiltert worden sind.

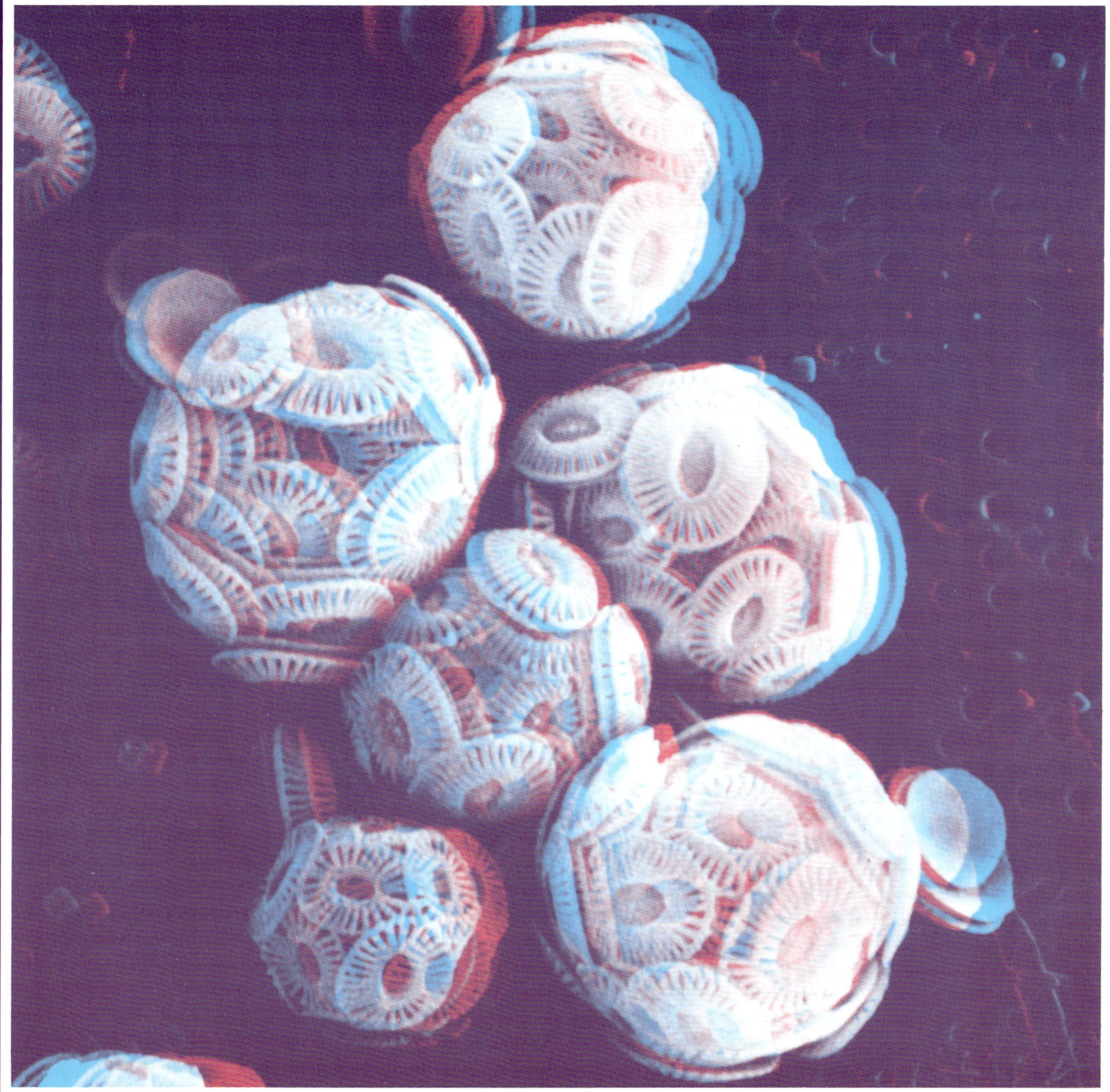

2-D: *5100X*↗; 3-D: *10200X*→

Plate 18: Coccolithophorid

These golden-brown algae evolved about 200 million years ago during the Jurassic Period, and their rapid evolutionary change through later periods produced many changes in their shapes and forms. Because of this, they are important in the determination of the age of ocean rocks and sediments. Many fine-grained limestones, such as the White Cliffs of Dover on the southeast channel coast of England, are made of coccoliths such as these.

Bild 18: Kalkalge

Diese goldbraune Alge entwickelte sich vor ungefähr 200 Millionen Jahre während der erdgeschichtlichen Formation des Juras. Während späterer Perioden unterlag sie starken Entwicklungen und änderte ihr Aussehen häufig. Aus diesem Grunde sind diese Kalkalgen ein wichtiges Hilfsmittel für die Altersbestimmung von Felsgestein und Ablagerungen in den Ozeanen. Viele feinkörnige Kalksteinfelsen, wie z.B. die weißen Felsen von Dover im Südosten der englischen Kanalküste sind aus solchen Kalkalgen aufgebaut.

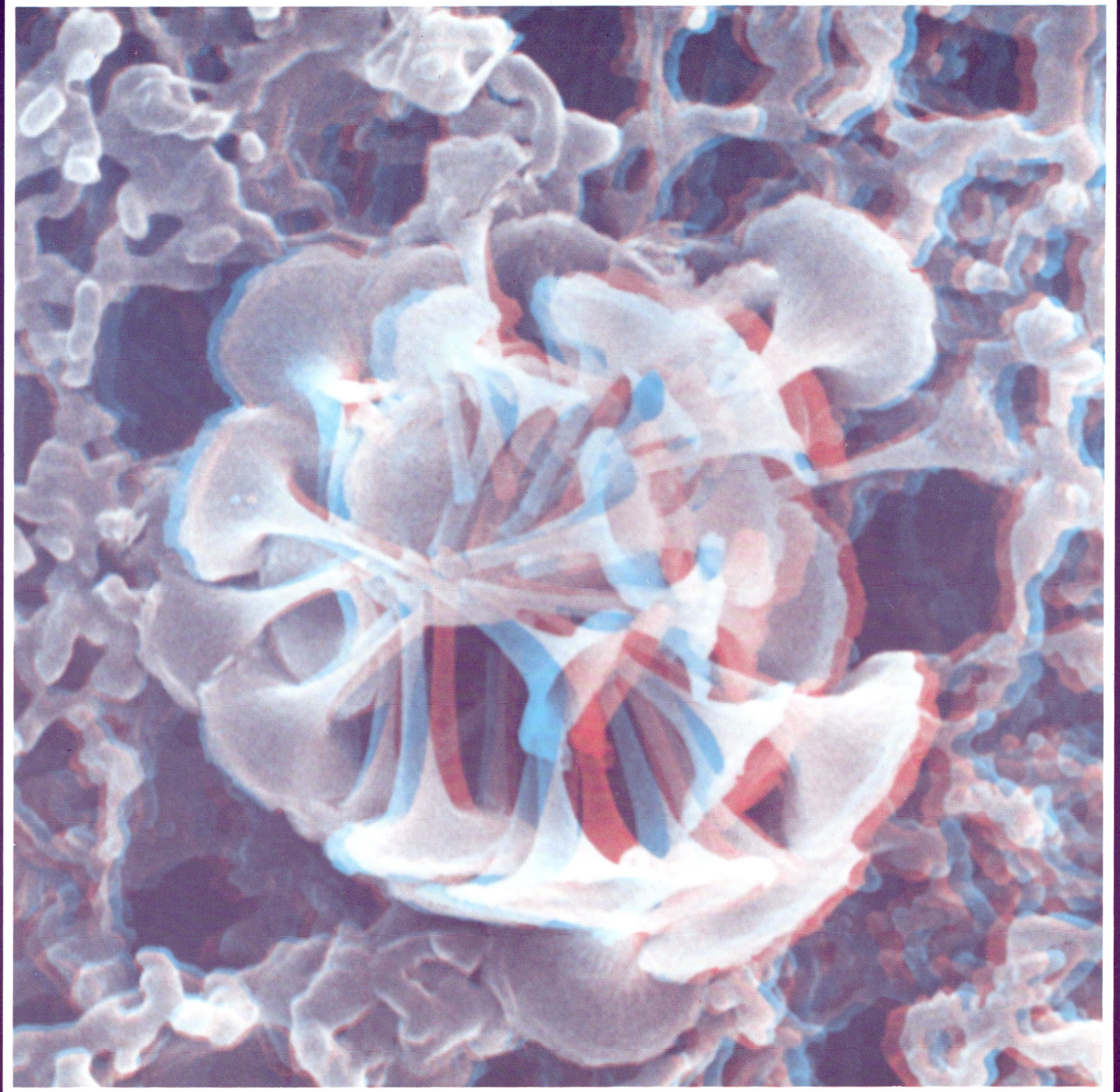

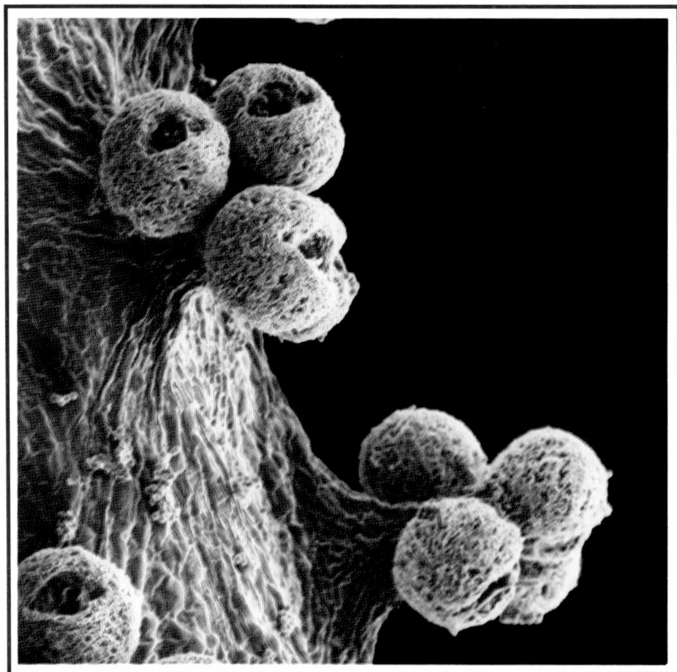

2-D: *50X*↗; 3-D: *100X*→

Plate 19: Pollen Sacs

These are the male "flowers" (raised pollen sacs) of a North American Jack-in-the-Pulpit plant, and you can see the tiny round pollen grains (one of which is shown further enlarged in Plate 20) within the openings of each sac. Because most of the grains never reach a flower, each plant produces tremendous numbers of them to insure that at least a few will reach other plants of their own kind and thus perpetuate their species. Pollen sacs similar to these have been found on fossils of a plant that lived 216 million years ago in the southwestern United States.

Bild 19: Pollensäcke

Das Bild zeigt die männlichen „Blüten" (Pollensäcke) einer in Nordamerika beheimateten Pflanze (Jack-in-the-Pulpit). Durch die Öffnungen erkennt man die winzigen, runden Pollenkörner im Inneren der Säkke. (Bild 20 zeigt ein solches Pollenkorn in stärkerer Vergrößerung.) Da die meisten Pollen verlorengehen und niemals eine Blüte erreichen, muß jede Pflanze riesige Mengen Pollen erzeugen, um sicherzustellen, daß wenigsten einige von ihnen eine andere Pflanze der gleichen Art erreichen und damit ihr Fortbestehen sichern. Ähnliche Pollensäcke wurden an einer versteinerten Pflanze gefunden, die vor 216 Millionen Jahren im Südwesten der Vereinigten Staaten gewachsen ist.

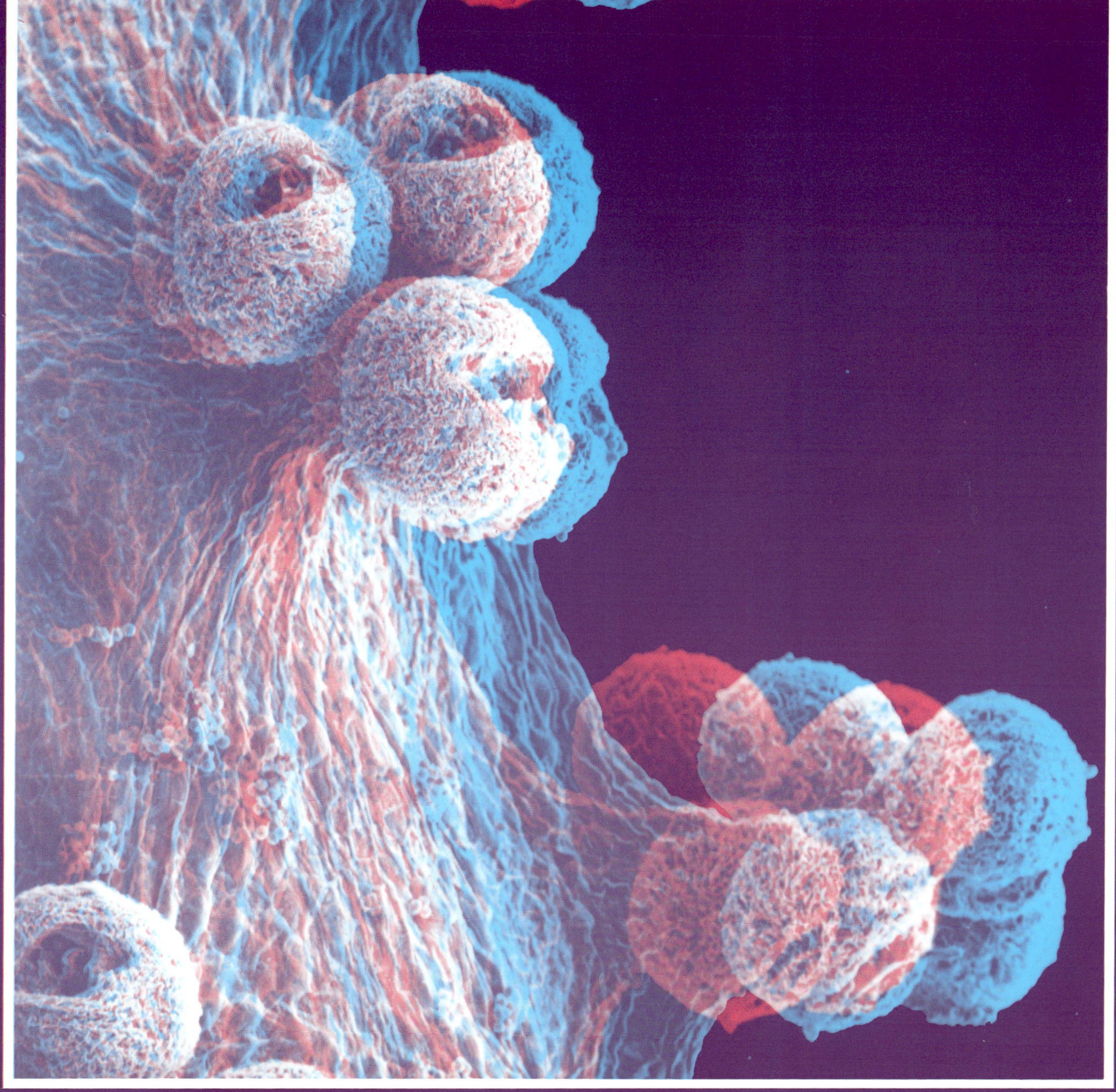

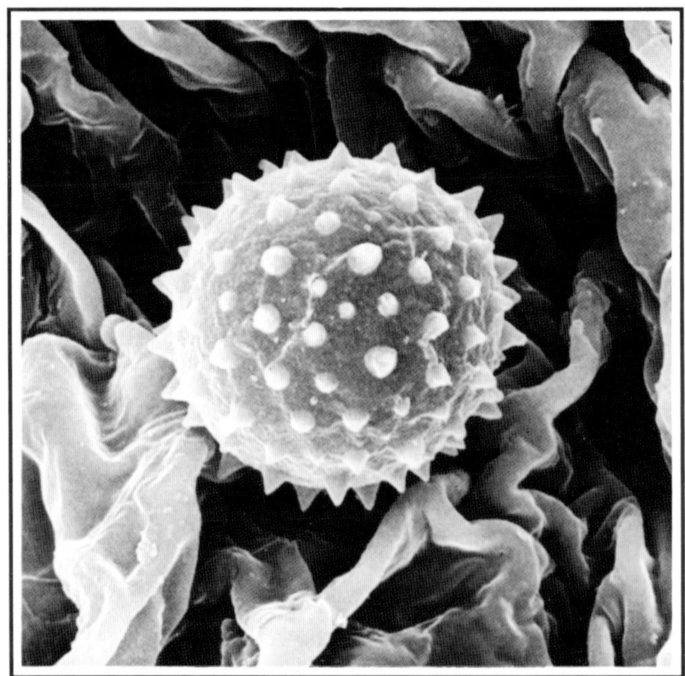

2-D: *2300X*↗; 3-D: *4600X*→

Plate 20: Pollen Grain

This single Jack-in-the-Pulpit pollen grain is still sitting on a fragment of the parent plant. Pollen grains such as this one are carried by wind and insects to transport sperm to fertilize other plants. The shapes, sizes, and other characteristics of the pollen of different plant species vary widely according to climate and method of transport; for example, pollen grains that are carried by air currents are very light. When a pollen grain lands on another plant of its kind, it grows a long tube down to the ovary below, through which it transfers genetic material to the egg inside.

Bild 20: Pollenkorn

Dieses Pollenkorn sitzt noch auf einem Fragment seiner Ursprungpflanze (Jack-in-the-Pulpit). Pollen wie diese werden vom Wind oder von Insekten davongetragen und transportieren Samen zur Befruchtung anderer Pflanzen. Form, Größe und andere Eigenschaften der Pollen unterschiedlicher Pflanzen variieren sehr stark und hängen vom Klima und der Transportmethode ab. So sind z.B. Pollen, die von Luftströmungen verbreitet werden, sehr leicht. Wenn ein Pollenkorn auf einer anderen Pflanze seiner Art landet, dann wächst ein langer Schlauch durch die Blüte zum Fruchtknoten, durch den genetisches Material zu den Samenanlagen übertragen wird.

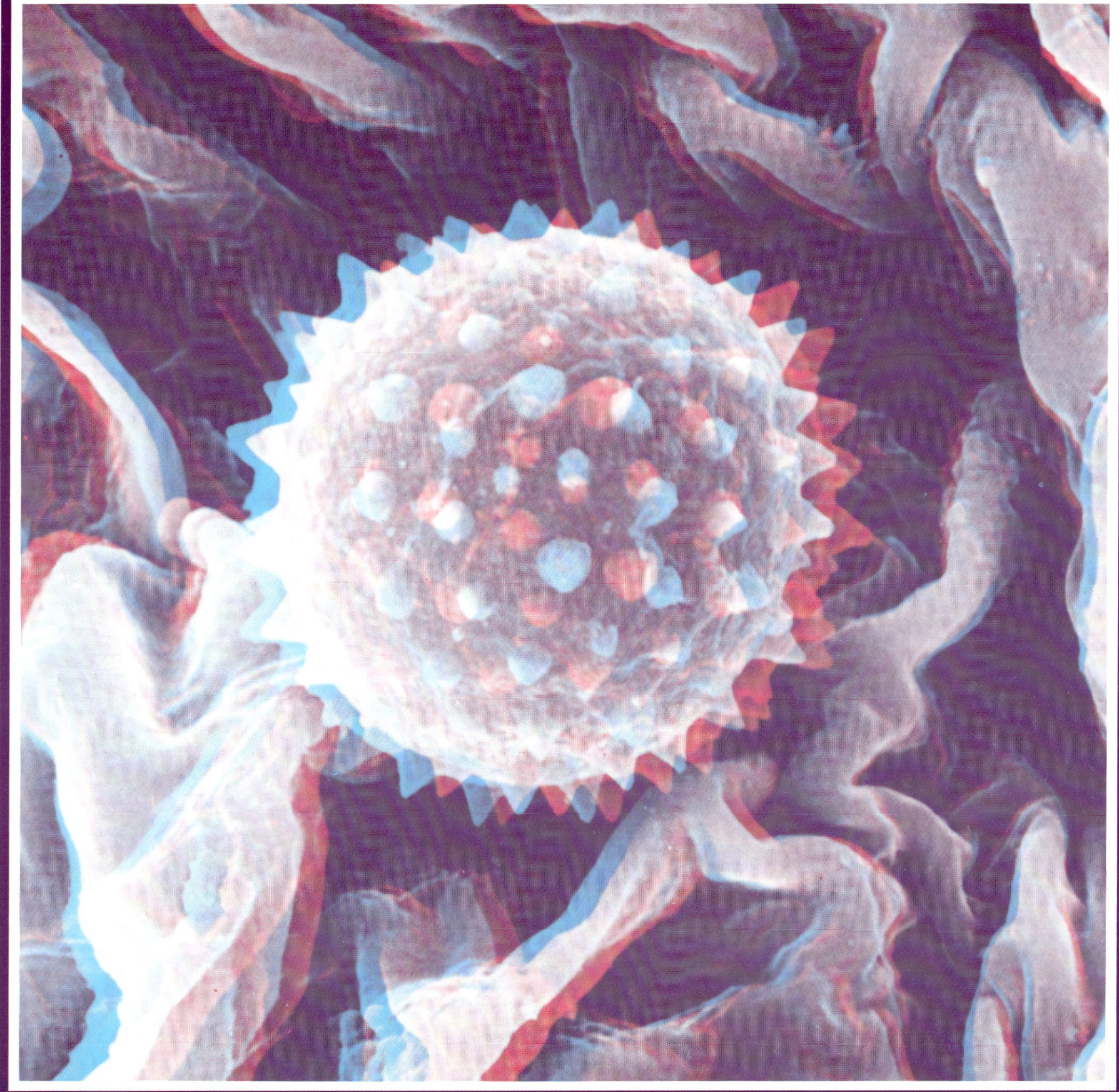

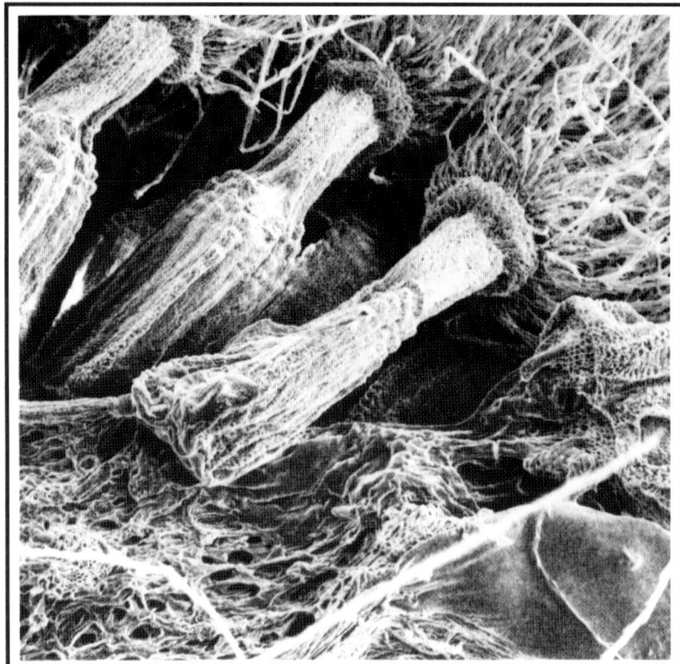

2-D: *55X*↗; 3-D: *110X*→

Plate 21: Dandelion Seeds

Seeds contain food stored for future plant embryos and are covered by one or two seed coats. They develop from ovules inside flowers that have been fertilized through a tube grown by a pollen grain (see Plate 20). In a fertilized seed, the process of germination creates an embryo, which then begins to grow into a young plant. For this micrograph, a dandelion blossom was cut open to reveal the row of seeds inside; in the natural course of events, though, the blossom would have opened of its own accord to allow air currents to carry the seeds away, borne on the wind by the many fluffy, parachute-like strands growing from their tops.

Bild 21: Samen des Löwenzahns

Samenkörner enthalten Nahrungsvorräte für ein Pflanzenembryo und sind von ein oder zwei Samenhäutchen überzogen. Sie entwickeln sich aus Eizellen im Inneren einer Blüte, die durch einen Pollenschlauch befruchtet worden sind (siehe Bild 20). Im befruchteten Samenkorn aktiviert der Vorgang des Keimens den Keimling, der dann zu einer jungen Pflanze heranwächst. Zum Zwecke dieser Aufnahme wurde eine Löwenzahnblüte aufgeschnitten, damit man die Samen sehen kann. Der natürliche Ablauf sieht selbstverständlich so aus, daß sich die Blüte von alleine öffnet, damit der Wind die Samen transportieren kann, getragen von einer Art Fallschirm aus vielen flauschigen Fäden, die aus ihrer Spitze herauswachsen.

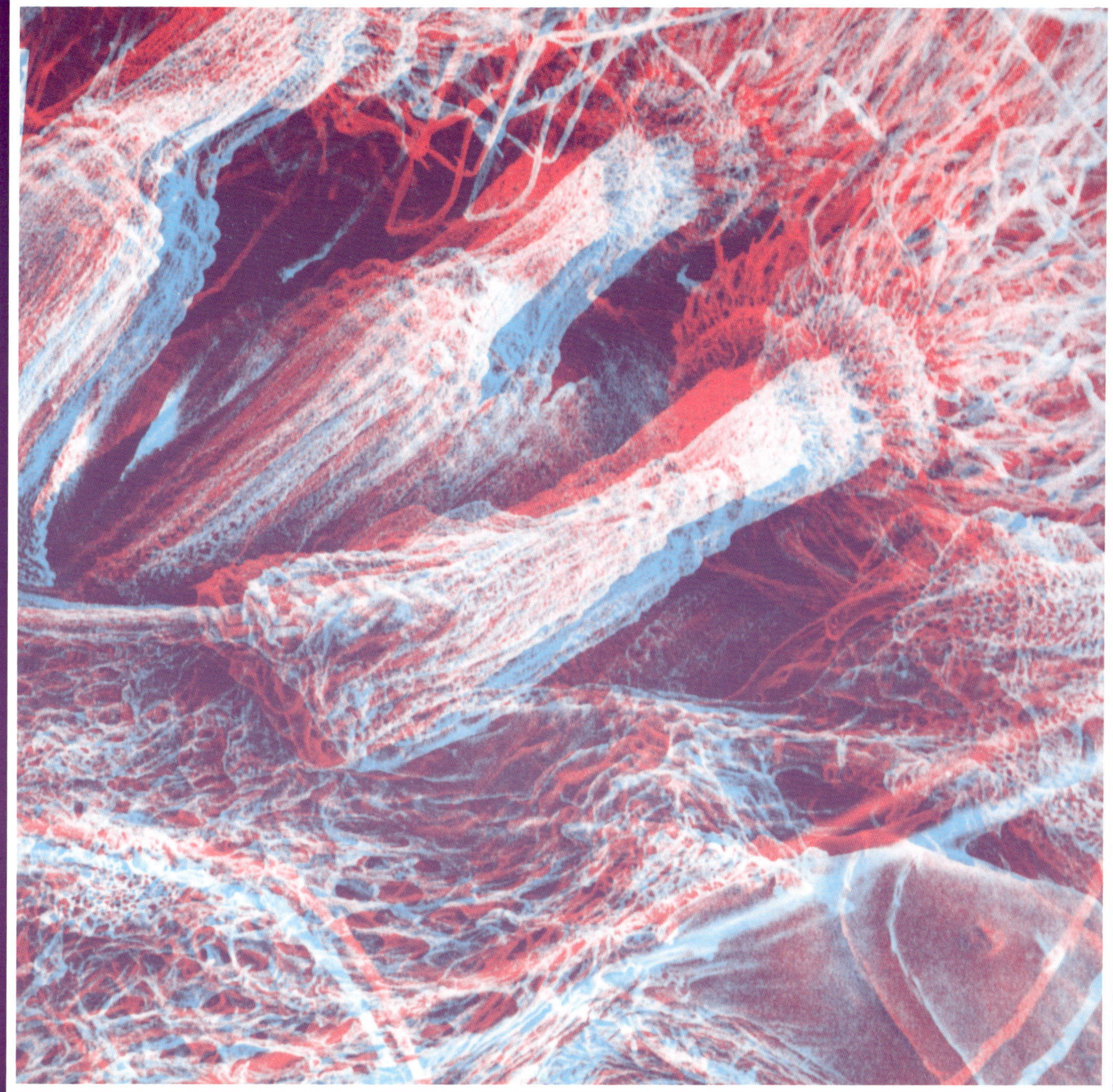

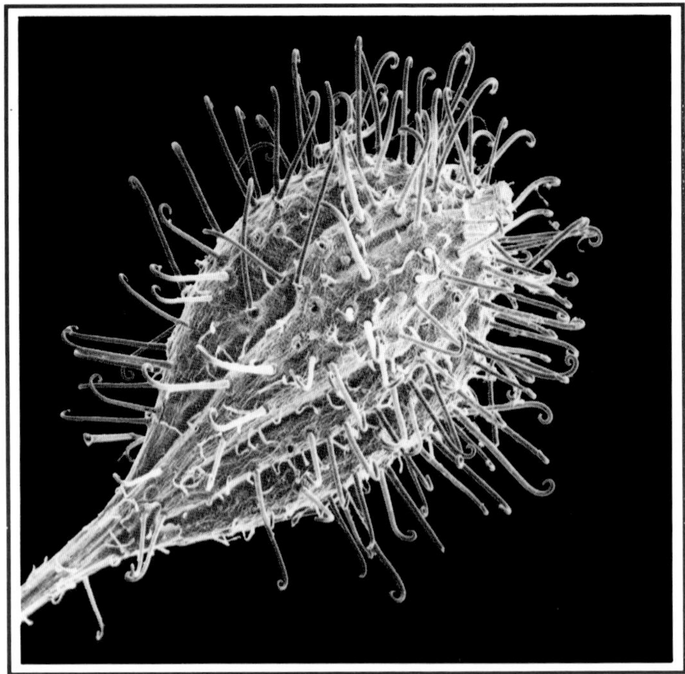

2-D: *18X*↗; 3-D: *36X*→

Plate 22: Enchanter's Nightshade Seed

Plants use a variety of methods to disperse their seeds to perpetuate their species. Some seeds are carried by the wind, some simply fall to the ground, and some are carried away by animals. The seeds of the Enchanter's Nightshade plant in West Virginia hitch a ride to another location by attaching themselves firmly to passing animals (including humans) by means of the many tiny hooks growing from their surface. It was the close examination of thistle blossom seeds very similar to this one that inspired the invention of Velcro (see Plate 5).

Bild 22: Samen des Enchanter's Nightshade

Mit ganz unterschiedlichen Methoden sorgen die Pflanzen für die Verbreitung ihrer Samen und damit für das Fortbestehen ihrer Art. Manche Samen werden vom Wind fortgetragen, andere fallen einfach zu Boden, und wieder andere werden von Tieren mitgenommen. Die Samen des Enchanter's Nightshade, einer Pflanze, die in West-Virginia, USA, vorkommt und die mit unserem Himmelsschlüsselchen eng verwandt ist, lassen sich wie Anhalter an einen anderen Ort bringen, indem sie sich mit Hilfe ihrer vielen, winzigen Häkchen fest an vorüberkommende Tiere und Menschen heften. Das genaue Studium des dem hier abgebildeten Samen sehr ähnlichen Distelsamens führte zur Erfindung des Klettverschlusses (Bild 5).

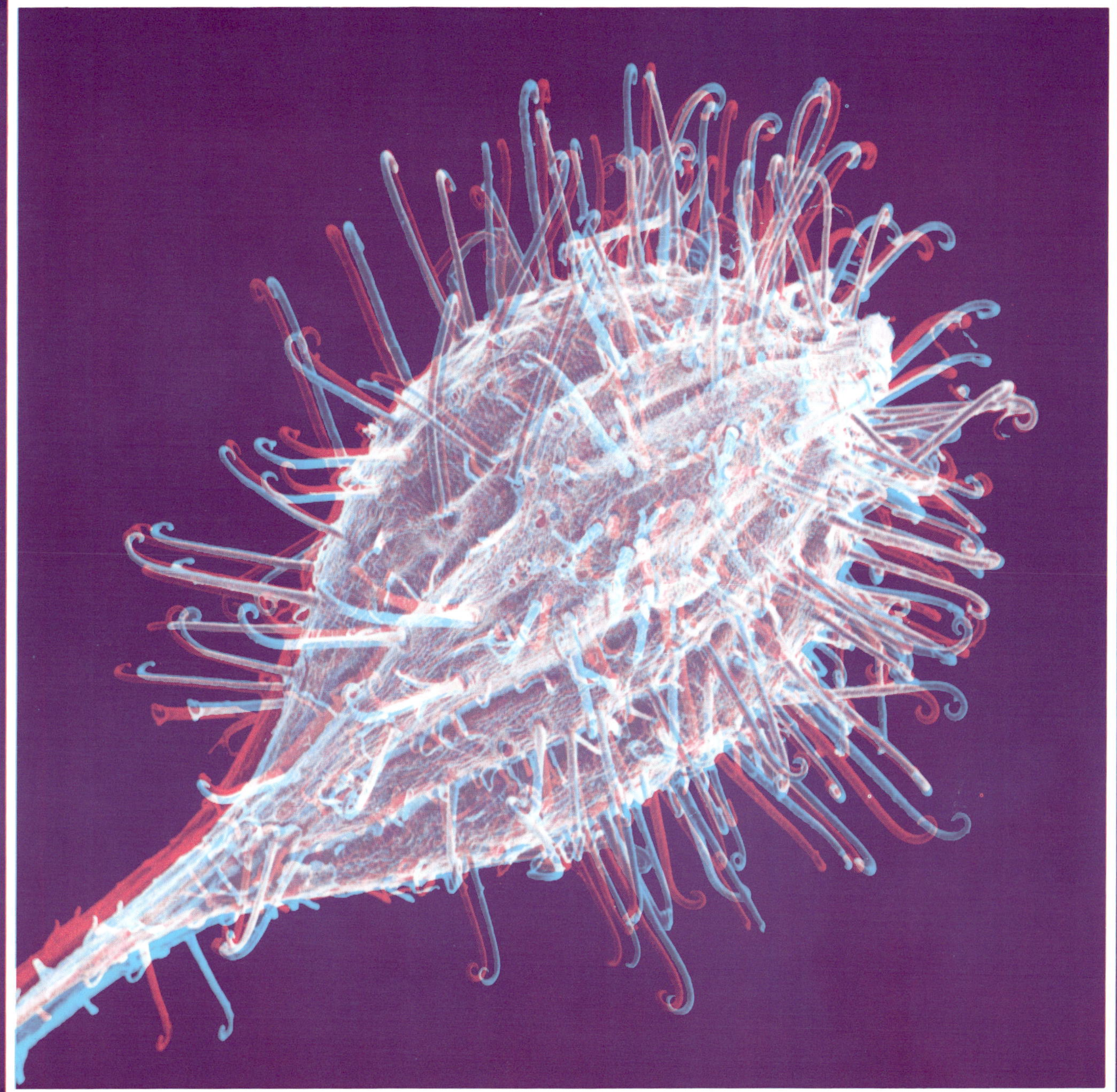

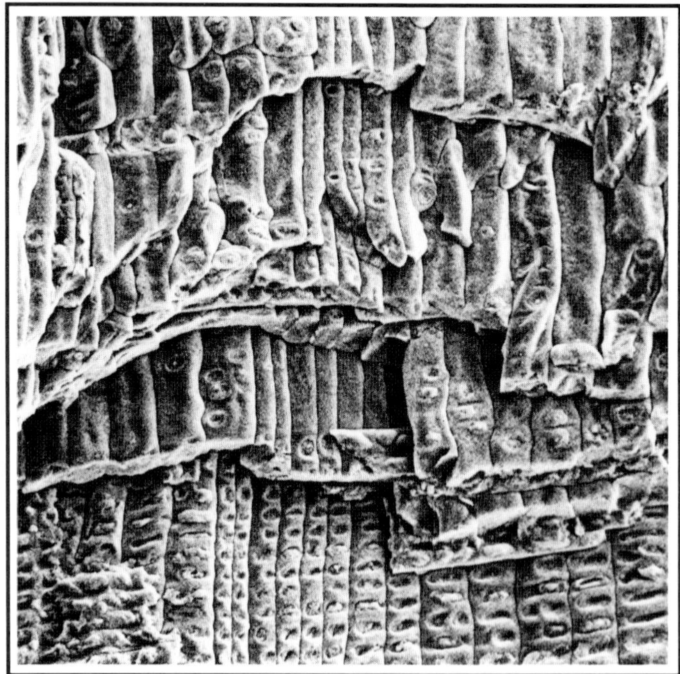

2-D: *160X*↗; 3-D: *320X*→

Plate 23: Petrified Wood

Some ancient plants and trees were carbonized over geologic time to become the vast oil and coal deposits we use for energy today. Others became petrified in a process that preserves once-living tissues by replacing them with minerals, which vary widely depending on the local environment. This tiny fragment of petrified wood came from an especially well-preserved fossilized conifer found in Tierra del Fuego at the southern tip of South America. Several layers of wood cells are shown here; you can see the pits that conducted sap from level to level in the living tree, much as the circulation system in animals transports blood to all parts of the body.

Bild 23: Versteinertes Holz

Urzeitliche Bäume und andere Pflanzen wurden im Verlaufe von geologischen Zeiträumen zu Kohle und Öl und bilden die Vorkommen, die wir heute für unsere Energieversorgung nutzen. Andere Pflanzen versteinerten in einem Vorgang, der ursprünglich lebende Gewebe erhält, indem sie durch Minerale ersetzt werden, abhängig von der jeweiligen Umgebung. Das winzige Stückchen versteinerten Holzes auf dem Bild stammt von einem besonders gut erhaltenen, fossilen Nadelbaum, der in der Tierra del Fuego in der südlichsten Spitze Süd-Amerikas gefunden wurde. Das Bild zeigt mehrere Schichten der Holzzellen. Man erkennt deutlich die Strahlen, in denen der lebende Baum die Säfte von Schicht zu Schicht leitete, ähnlich dem Kreislaufsystem der Lebewesen, das Blut zu allen Teilen des Körpers befördert.

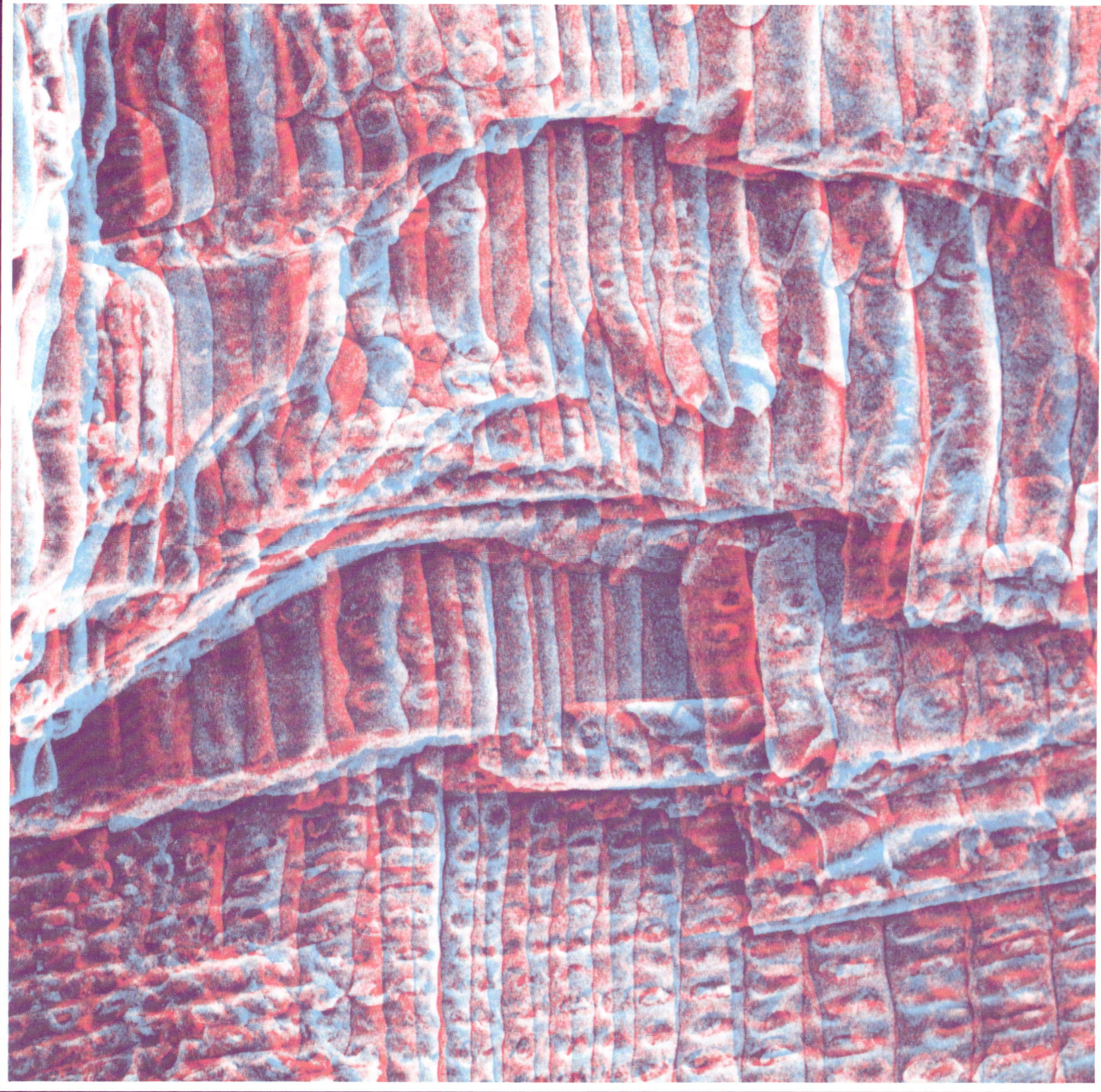

2-D: *130X*↗; 3-D: *260X*→

Plate 24: Table Salt

Salt, a common household item, is made up of the elements sodium and chlorine (each of which is quite poisonous by itself), which combine to form the cube-shaped crystals of sodium chloride you see in this micrograph. In some parts of the world, the scarcity of salt has made it an important trade commodity of high value, and wars have even been fought over the possession of important salt deposits. We think of salt primarily as a seasoning for foods, but its uses in many biological, chemical, and industrial processes are countless.

Bild 24: Kochsalz

Das Salz, das in jedem Haushalt verwendet wird, besteht aus den chemischen Elementen Natrium und Chlor. (Jedes dieser Elemente ist für sich allein sehr giftig.) Die würfelförmigen Kristalle des Kochsalzes zeigt das Bild. In manchen Teilen der Welt war Kochsalz selten und daher ein wichtiges und wertvolles Handelsgut, um Salzvorkommen wurden sogar Kriege gefürht. Beim Wort „Kochsalz” denkt man zuerst an die Verwendung bei der Zubereitung von Speisen, aber auch die Anwendungen in biologischen, chemischen und industriellen Prozessen sind zahllos.

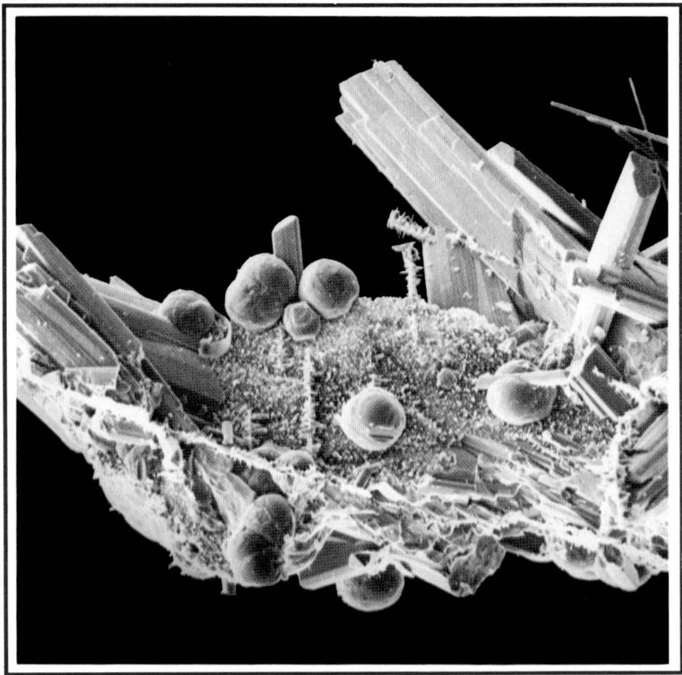

2-D: *140X*↗; 3-D: *280X*→

Plate 25: Franklin Micromineral

The neighboring Franklin and Sterling Hill deposits in New Jersey together form the richest site of mineral species known in the world. Over 350 mineral species have been found so far, some of which occur only there and nowhere else. A number of these species comprise the most spectacular collection of fluorescent minerals known. These ore bodies were deposited about one billion years ago and were part of an ancient rift valley that has since seen a lot of geologic activity. Five periods of mountain building and several periods of volcanism have all contributed to its unique variety of minerals.

Bild 25: Franklin-Mikromineral

Die reichsten Mineralienvorkommen der Welt sind die benachbarten Franklin und Sterling Hill in New Jersey, USA. Mehr als 350 verschiedene Mineralien sind bis jetzt hier gefunden worden. Einige von ihnen treten nur hier auf und sonst nirgendwo. Einige dieser Arten bilden die bekannteste Ansammlung von fluoreszierenden Mineralien. Die Vorkommen bildeten sich vor etwa einer Milliarde Jahren und waren Teil eines urzeitlichen Grabenbruchs, der seitdem viele weitere geologische Aktivitäten erlebt hat, unter anderem Zeitalter, in denen sich Berge bildeten, oder Perioden des Vulkanismus. Alle diese Aktivtäten haben zur Bildung dieses einzigartigen Reichtums an Mineralier beigetragen.

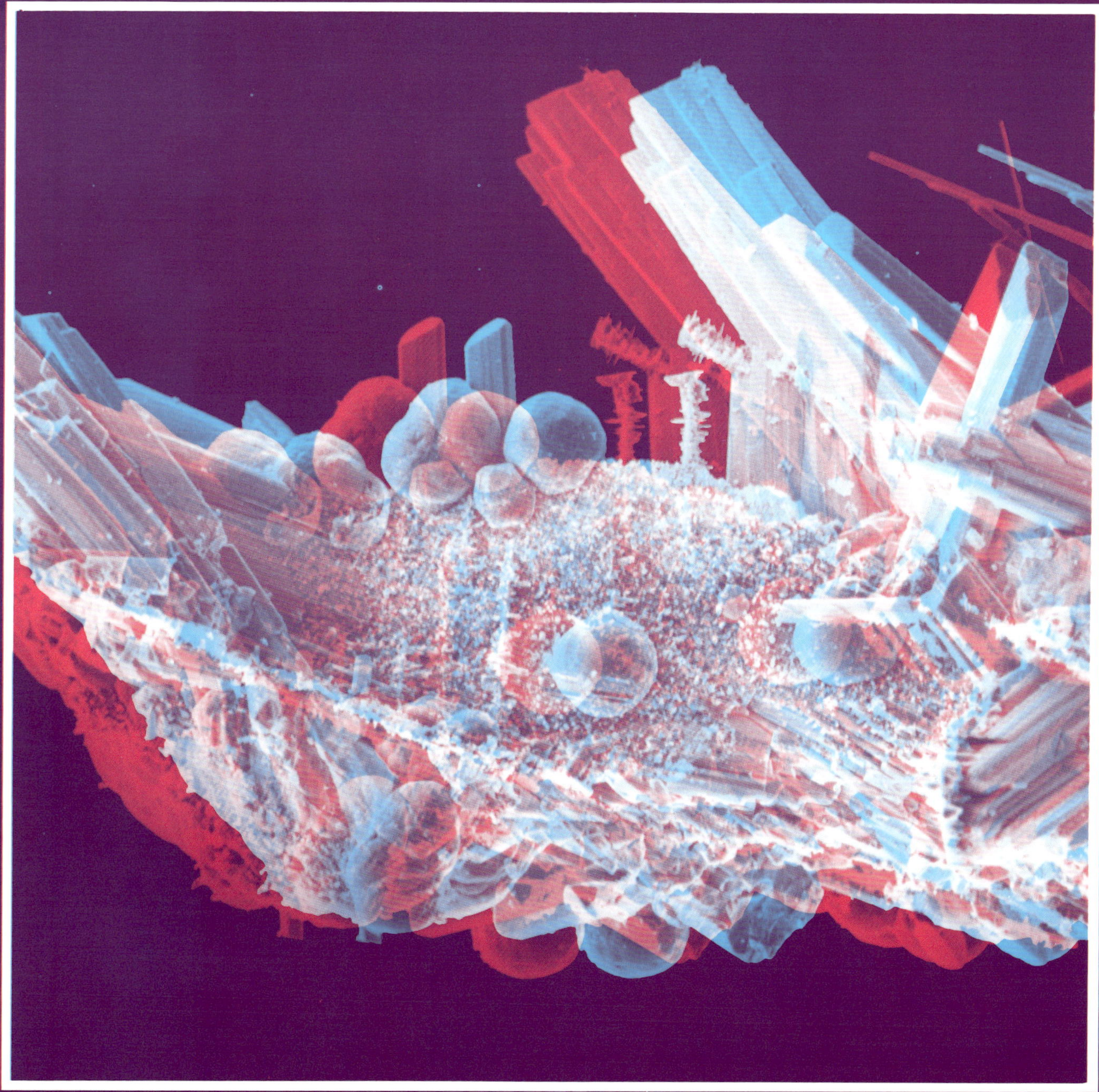

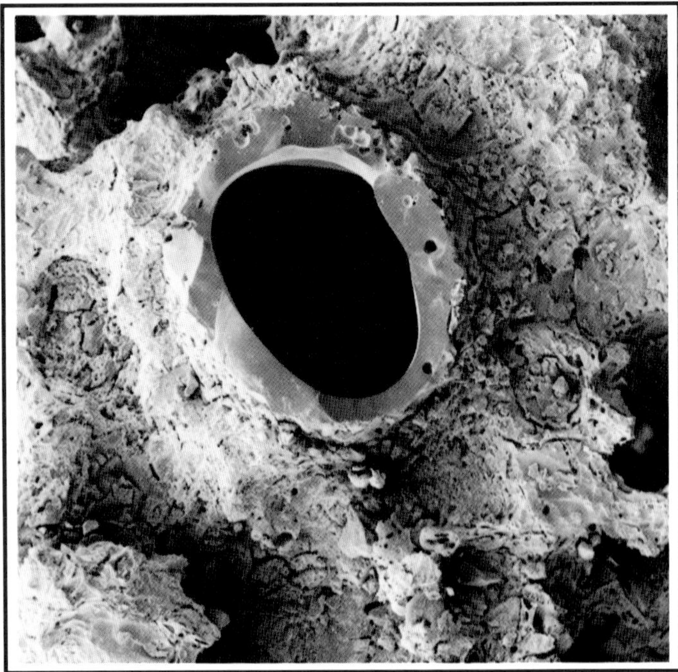

2-D: *22X*↗; 3-D: *44X*→

Plate 26: "Petrified" Lightning

Lightning bolts strike with incredible speed and produce intense heat. When a bolt strikes sand, it bores through the quartz mineral in less than one ten-thousandth of a second, melting the grains of sand. This forms a hollow tube surrounded by fulgurite, the mineral created when lightning strikes and fuses sand. Here you see the tiny, fragile glass tube created by a microscopic lightning bolt that struck a sand hill in North Carolina. It's surrounded by intact sand grains that were far enough from the path of the bolt to remain unmelted.

Bild 26: „Versteinerter" Blitz

Blitze schlagen mit unvorstellbarer Geschwindigkeit ein und erzeugen hohe Temperaturen. Wenn ein Blitz in Sand einschlägt, bohrt er sich in weniger als einer Zehntausendstel Sekunde einen Weg durch den Quarz und bringt die Sandkörner zum Schmelzen. Dabei bildet sich eine hohle Röhre („Blitzröhre") aus Fulgurit, dem Mineral, das beim Blitzschlag aus Quarz entsteht. Das Bild zeigt die winzige, zerbrechliche Röhre, die entstand, als ein mikroskopisch kleiner Blitz in einen Sandhügel einschlug. Die Röhre ist umgeben von unzerstörten Sandkörner, die vom Pfad des Blitzes weit genug entfernt waren und nicht geschmolzen sind.

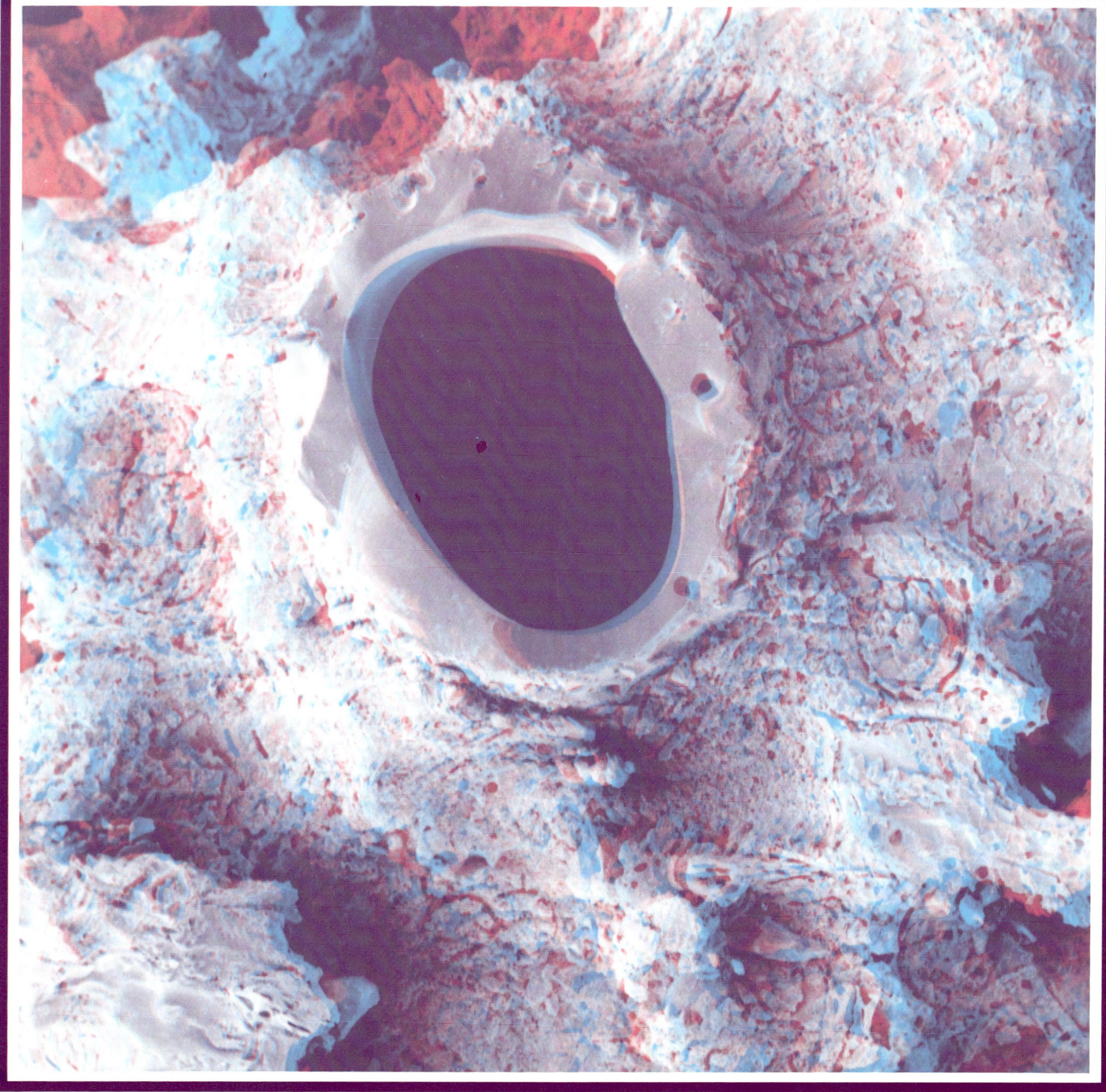

2-D: *1700X*↗; 3-D: *3400X*→

Plate 27: Pyroxene Crystal

Rocks of the earth's mantle exist at great depths and are seen at its surface only in a few places where tektonic forces propelled them upward ages ago. Zabargad, a small desert island in the Red Sea near Egypt, is one such place. Originally a block of the earth's mantle lying more than 60 miles beneath the ocean floor, it was lifted to the surface by the tremendous forces that separated Africa from the Arabian Peninsula, resulting in the creation of the Red Sea. Shown here is a single crystal face of the mineral pyroxene in a peridotite rock that is nearly pure mantle material. Peridot gems were found extensively on Zabargad and mined there since ancient times.

Bild 27: Pyroxene-Kristall

Das Gestein des Erdmantels reicht bis in große Tiefen. Tiefengestein ist nur an wenigen Stellen der Erdoberfläche sichtbar, nämlich da, wo es vor Urzeiten von starken tektonischen Kräften an die Oberfläche gedrückt wurde. Zabarjat, eine kleine Wüsteninsel im Roten Meer ist ein solcher Ort. Ursprünglich war diese Insel ein Teil des Erdmantels und lag 100 km unterhalb des Meeresbodens. Die Insel entstand, als infolge eines ungeheuren Drucks die Arabische Halbinsel von Afrika getrennt wurde und dabei das Rote Meer entstand. Das Bild zeigt eine Pyroxene-Einkristallfläche in Peridotit-Felsgestein, dem Material, aus dem der Erdmantel nahezu vollständig besteht. Peridot-Schmucksteine sind auf der Insel Zabarjat zu finden und wurden dort schon in der Antike geschürft.

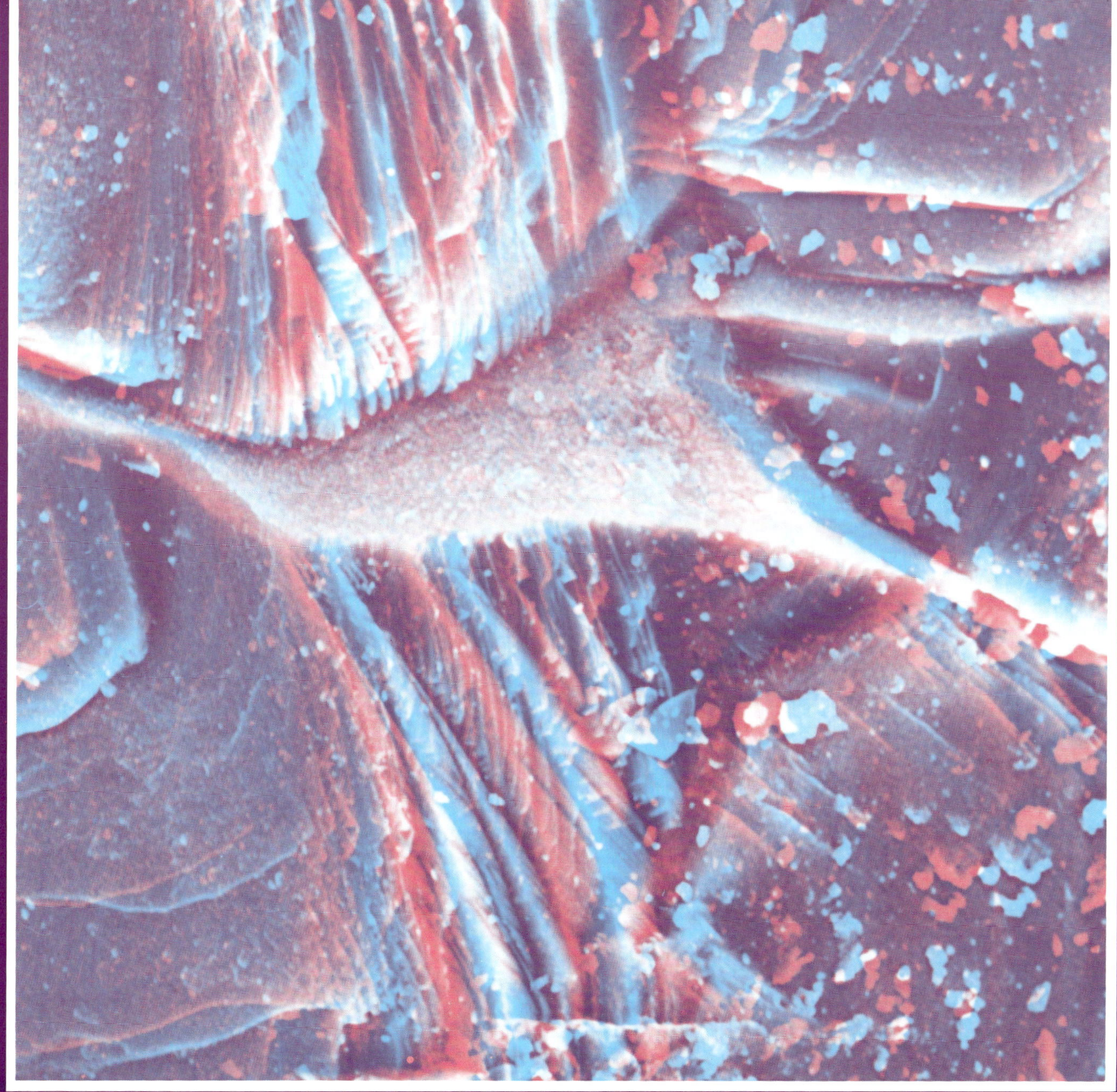

2-D: *420X*↗; 3-D: *840X*→

Plate 28: Volcanic Ash

Volcanic eruptions occur when molten rock from the earth's hot interior rises to the surface and is ejected into air or water. The glassy particles in this micrograph are from a gigantic eruption about 10 million years ago on the North American continent in what is now the state of Idaho. The resulting ash-fall was responsible for the death of a group of Miocene Age animals, including rhinos, camels, horses, and birds, that had gathered around a water hole in present-day Nebraska. They died from breathing the falling ash, which then buried them so quickly that they remained unusually well preserved.

Bild 28: Vulkanasche

Ein Vulkan bricht aus, wenn geschmolzenes Gestein aus dem heißen Erdinneren an die Oberfläche steigt und in die Luft oder ins Wasser ausgestoßen wird. Die glasigen Partikel auf der Abbildung stammen von einer gigantischen Eruption, die sich vor 10 Millionen Jahren in Nordamerika in der Gegend des heutigen Idaho ereignet hat. Der dabei aufgetretene Ascheregen war verantwortlich für den Tod einer Gruppe von Tieren dieses Zeitalters (Miozän), unter anderem Rhinozerosse, Kamele, Pferde und Vögel, die sich um eine Wasserstelle im heutigen Nebraska gesammelt hatten. Todesursache war das Einatmen der Asche, die sie dann auch so schnell unter sich begrub, daß sie alle ungewöhnlich gut erhalten blieben.

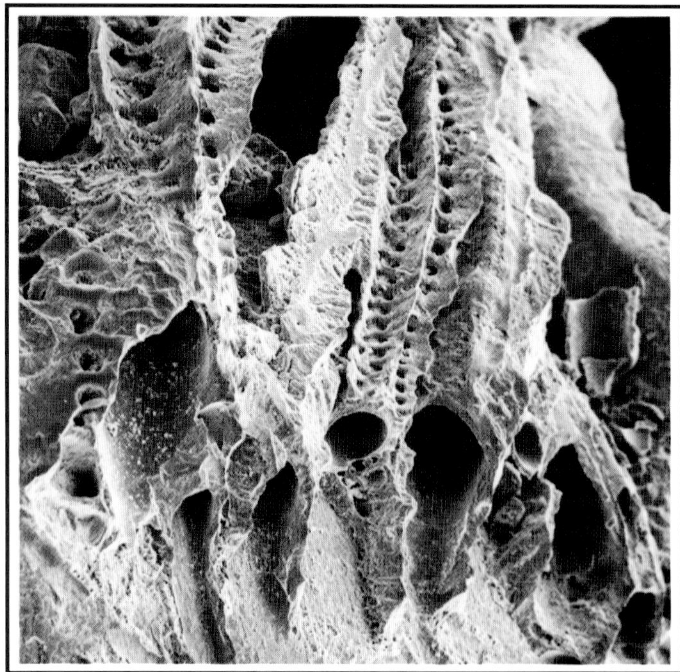

2-D: *60X*↗; 3-D: *120X*→

Plate 29: Worm Tubes

Polychetes are segmented sea worms that feed on small particles floating in the water. Many types of these animals are tube-dwellers that build on hard surfaces such as corals, rocks, and shells; if enough tubes are built in one place, they form a crust, as shown in this micrograph of a piece of a shell found in the Gulf of California. Each tube was formed by an individual worm, and you can often find several very different worm species piled on top of each other, particularly if the rock or shell upon which they've settled is in an area of strong currents.

Bild 29: Wurmröhren

Polychäten sind meerbewohnende Borstenwürmer, die von Schwebestoffen im Wasser leben. Viele Arten dieser Würmer leben in Röhren, die sie auf harten Oberflächen, wie Felsen, Korallen und Muscheln bauen. Viele dieser Röhren zusammen bilden eine Art Kruste, wie auf diesem Bild, das ein Stück einer Muschel aus dem Golf von Kalifornien darstellt. Jede Röhre wurde von einem anderen Wurm gebaut, und oft findet man ganz unterschiedliche Arten von Würmern aufeinander, besonders, wenn der Felsen oder die Muschel, die sie besiedeln, sich in einer starken Strömung befindet.

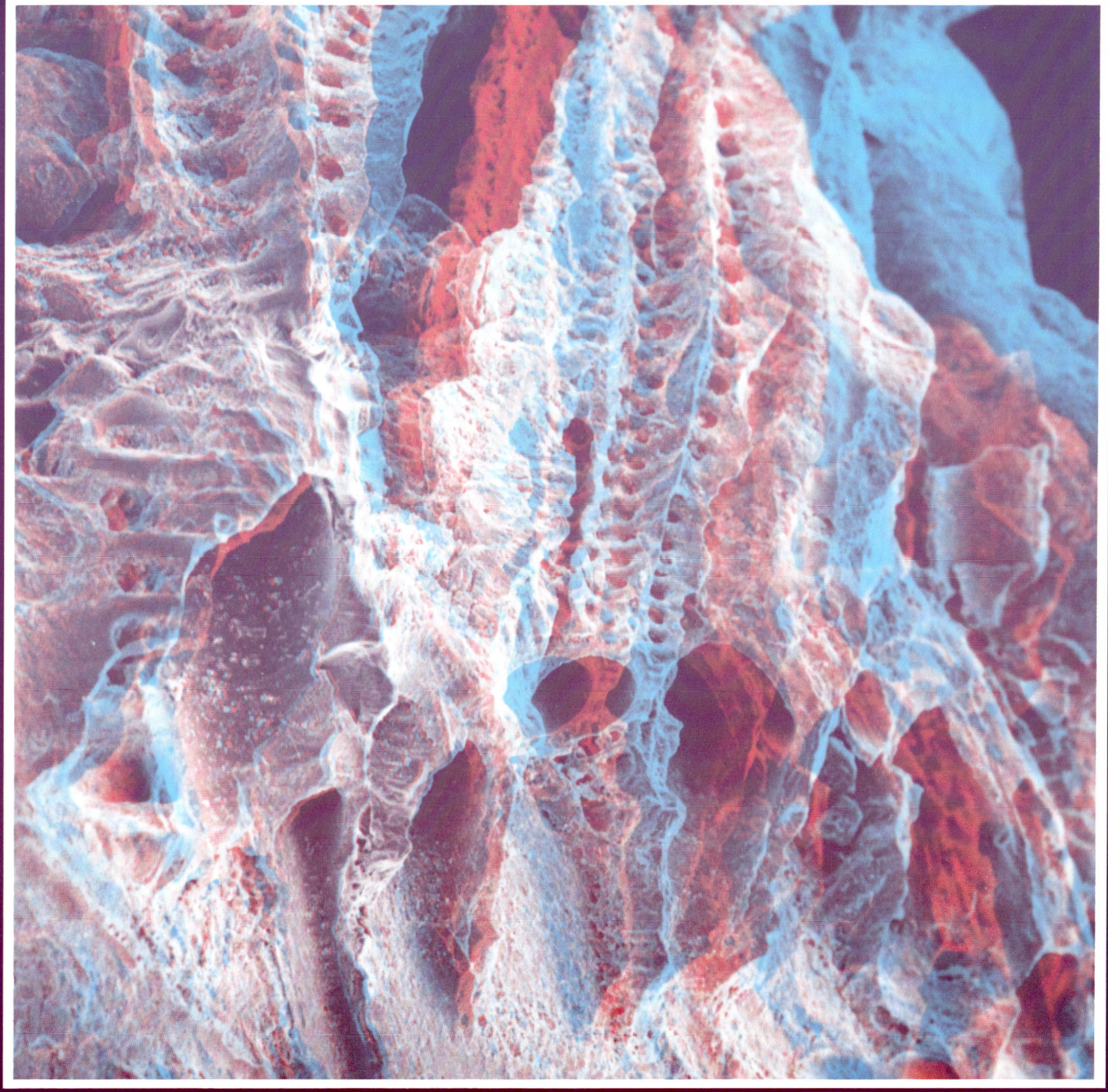

2-D: *550X*↗; 3-D: *1100X*→

Plate 30: Burned Tungsten Wire

If you attach a short piece of tungsten wire between two electrodes and send an electrical current through it until it flames – much as when a light bulb burns out – you'll see something like this in your SEM. These dendritic crystals grew during condensation of the vapor phase after the tungsten burned, and they're similar to the ice crystals that form on windows during a sudden winter chill. These crystals have grown outward in all directions of the third dimension, and they're so tiny that only an instrument capable of high magnification and resolution can show their intricate forms.

Bild 30: Durchgebrannter Wolframdraht

Wenn man ein Stück Wolframdraht zwischen zwei Elektroden ausspannt und Strom hindurchschickt, bis er aufflammt – vergleichbar einer Glühbirne, die durchbrennt – erhält man das, was man auf diesem Bild sieht: Die verästelten Kristalle entstanden während des Niederschlags des Wolframdampfes nach dem Verbrennen. Sie ähneln den Eiskristallen, die sich bei einem plötzlichen Frosteinbruch an Fensterscheiben bilden. Die Kristalle hier wachsen in alle Richtungen, und sie sind so winzig, daß nur ein Mikroskop mit hoher Vergrößerung und guter Auflösung ihren komplizierten Aufbau wiedergeben kann.

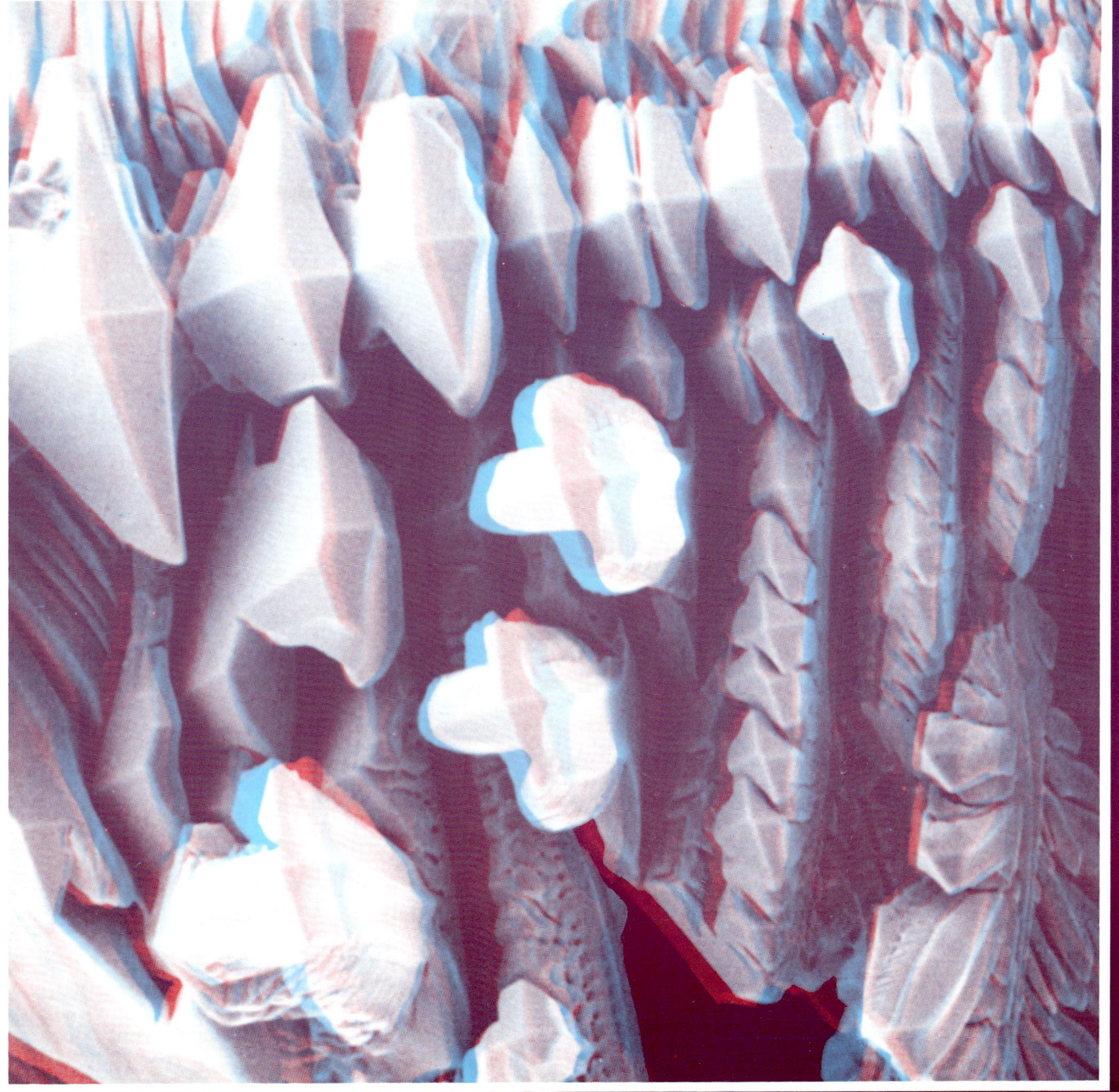

ACKNOWLEDGEMENTS

The technical foundation of this work has come from many sources. I am particularly grateful to Leica, Inc. and the Polaroid Corporation, Jeff Streger of Leica/Cambridge and Marcel Broussard of OSC Technologies (two excellent service engineers), and Dick Daniel of Radco. Many of the specimens were provided through the generous donations of scientists, medical researchers, graduate students, and a few special friends. I wish to thank the following contributors who donated their samples or expertise, and often both, to the plates published in this work:

DANKSAGUNG

Zum technischen Gelingen dieser Arbeit haben Viele beigetragen. Mein besonderer Dank gilt den Firma Leica, Inc. und Polaroid Corporation, außerdem Jeff Streger von der Firma Leica/Cambridge und Marcel Broussard von der Firma OSC Technologies (beide sind hervorragende Service-Ingenieure) sowie Dick Daniel von der Firma Radco. Viele der Proben verdanke ich der Großzügigkeit von Wissenschaftlern, Medizinern, Graduierten und einigen besonderen Freunden. Im einzelnen möchte ich den folgenden Damen und Herren danken, die mir Proben oder Hintergrund-Informationen oder beides zu den einzelnen Bildern diese Buches überlassen haben:

1: William A. Newman, Professor of Biological Oceanography, Scripps Institute of Oceanography, La Jolla, Calif.

2: Mahto Topah, Saugerties, N.Y.

4: Mike Jones, Jersey City, N.J.

8, 9: Paul LoGerfo, Prof. of Surgery and Dir. of Surgical Oncology, Columbia-Presbyterian Medical Center, New York, N.Y.

10: C.S. Pitchumani, Professor of Medicine/Community and Preventative Medicine, New York Medical College, and Chief of the Division of Gastroenterology, Our Lady of Mercy Medical Center, Bronx, N.Y.

11: Marianne Wolfe, Professor of Pathology, Columbia-Presbyterian Medical Center, New York, N.Y. and Attending Pathologist, Morristown Memorial Hospital, Morristown, N.J., Jerold Brett, Senior Staff Associate, and Steven Brunnert, D.V.M., Assistant Professor, Department of Pathology, both at Columbia-Presbyterian Medical Center, New York, N.Y.

12, 13: Betty Faber, Staff Scientist, Liberty Science Center, Jersey City, N.J.

14: Kenneth G. Miller, Prof. of Geological Sciences, Rutgers University, Piscataway, N.J. and Adjunct Senior Scientist, LDEO.

15, 16: Richard Mortlock, Senior Staff Associate, LDEO, and O. Roger Anderson, Professor, Teachers College of Columbia University and Senior Research Scientist, LDEO.

17, 18: Andrew McIntyre, Senior Research Scientist, LDEO and Professor of Geology, Queens College, City University of New York, Flushing, N.Y.

19, 20: Bruce Cornet, former Research Associate, LDEO, and Sarah J. Fowell, Research Associate, University of South Carolina, Columbia, S.C.

22: Dorothy M. Peteet, Research Scientist, Goddard Institute for Space Studies, New York, N.Y. and Adjunct Research Scientist, LDEO.

23: Blanca and Ricardo Covasevich, Tierra del Fuego, Chile, and Gordon C. Jacoby, Senior Research Scientist, Tree Ring Laboratory, LDEO.

25: Herb Yeates, Franklin Ogdensburg Mineralogical Soc., and John Baum, Curator, Franklin Mineral Museum, Franklin, N.J.

27: Enrico Bonatti, Doherty Senior Scientist, LDEO.

28: Janet Salzman, Graduate Research Assistant, LDEO.

29: Nicky Huard, Montreal, Canada, and Kristian Fauchauld, Department of Invertebrate Zoology, National Museum of Natural History, Smithsonian Institution, Washington, D.C.

30: Siu-Wai Chan, Associate Prof. of Materials Science, Henry Krumb School of Mines, Columbia University, New York, N.Y.

Discussions of some of the micrographs in this book have been paraphrased from various public sources, but the following specific books and articles can provide further information for the interested reader:

Bei der Abfassung der Kommentare zu den einzelnen Bildern habe ich auf zahlreiche Quellen zurückgegriffen. Die folgenden Bücher und Artikel dürften dem interessierten Leser weitere Informationen zur Verfügung stellen:

Tissues and Organs: a Text Atlas of Scanning Electron Microscopy by R.D. Kessel and R.H. Kardon (W.H. Freeman and Co., San Francisco, Calif., 1979).

Introduction to Marine Micropaleontology, B.U. Haq and A. Boersma, eds. (Elsevier Press, 1978).

"How an Eggshell is Made" by T.G. Taylor, in *Scientific American*, March 1970.